COMMUNICATION AND COMMUNITY

Science and Psychoanalysis. Volume VIII.

COMMUNICATION AND COMMUNITY

Edited by

JULES H. MASSERMAN, M.D.
Professor and Co-Chairman of Psychiatry
Northwestern University
Chicago

GRUNE & STRATTON

New York · London · 1965

Preceding volumes of *Science and Psychoanalysis*

Volume I: *Integrative Studies* (1958)
Volume II: *Individual Familial Dynamics* (1959)
Volume III: *Psychoanalysis and Human Values* (1960)
Volume IV: *Psychoanalysis and Social Process* (1961)
Volume V: *Psychoanalytic Education* (1962)
Volume VI: *Violence and War* (1963)
Volume VII: *Development and Research* (1964)

Library of Congress Catalog Number 58-8009

GRUNE & STRATTON, INC.
381 Park Avenue South
New York 16, New York

Printed and bound in the United States of America (K-B)

CONTENTS

Techniques and Services

CONTRIBUTORS

ADLER, LETA MCKINNEY, PH.D.: Assistant Clinical Professor (Sociology) Department of Psychiatry, University of Southern California School of Medicine, Los Angeles, California.

ARIETI, SILVANO, M.D.: William A. White Institute, New York, New York.

BLOCH, DONALD, A., M.D.: Director of Research, Jewish Board of Guardians, New York, New York.

BRYT, ALBERT, M.D.: William A. White Institute, New York, New York.

CHRZANOWSKI, GERARD, M.D.: Training Analyst, William A. White Institute, New York, New York.

DATTA, LOIS-ELLIN, PH.D.: Laboratory of Psychology, National Institute of Mental Health, Washington, D.C.

DUHL, LEONARD, J., M.D.: Chief, Office of Planning, National Institute of Mental Health, Washington, D. C.

ECKARDT, MARIANNE HORNEY, M.D.: Washington Psychoanalytic Institute, Washington, D.C.

ELKES, JOEL, M.D.: Henry Phipps, Professor and Director, Department of Psychiatry, The Johns Hopkins Hospital, Baltimore, Maryland.

ENELOW, ALLEN J., M.D.: Associate Clinical Professor, University of Southern California School of Medicine, Los Angeles, California.

GELB, LESTER A., M.D.: Instructor in Psychiatry, Columbia University, New York, New York.

GELLER, JOSEPH J., M.D.: William A. White Institute, New York, New York.

HANFMANN, EUGENIA, PH.D.: Psychological Counselling Center, Brandeis University, Waltham, Massachusetts.

HARRIS, IRVING D., M.D.: Associate Professor of Clinical Psychiatry, University of Illinois, Chicago, Illinois.

HEATH, ROBERT G., M.D., D.M.Sc.: Professor and Chairman, Department of Psychiatry and Neurology, Tulane University School of Medicine, New Orleans, Louisiana.

HOCH, PAUL H., M.D.: Formerly Commissioner of Mental Hygiene, State of New York, Albany, New York. (Deceased, Dec. 15, 1964)

LAWSON, ELENA PADILLA, PH.D.: Postgraduate Center for Mental Health, New York, New York.

LEOPOLD, ROBERT L., M.D.: Assistant Professor of Psychiatry, University of Pennsylvania Medical School, Philadelphia, Pennsylvania.

MAZER, MILTON, M.D.: William A. White Institute, New York, New York.

MILLER, JAMES G., M.D., PH.D.: Director, Mental Health Research Institute, Ann Arbor, Michigan.

MILLET, JOHN, A. P., M.D.: Training Psychoanalyst, New York University School of Psychiatry, New York, New York.

PARLOFF, MORRIS B., PH.D.: Laboratory of Psychology, National Institute of Mental Health, Washington, D.C.

SCHACHTER, SHEMAN O., M.D.: Medical Director, Hudson Guild Counseling Center, New York, New York.

SHANDS, HARLEY C., M.D.: Professor of Psychiatry, Downstate Medical Center, State University of New York, Brooklyn, New York.

SOBEL, RAYMOND, M.D.: Associate Professor of Psychiatry and Pediatrics; Head, Division of Child Psychiatry, University of Washington School of Medicine, Seattle, Washington.

SPIEGEL, ROSE, M.D.: Associate Professor of Psychiatry, New York School of Psychiatry, New York, New York.

TARAIL, MARK, D. S. W.: Staff, Maimonides Hospital, Brooklyn, New York.

TULIPAN, ALAN B., M.D.: Medical Director, The Guidance Center, New Rochelle, New York.

ULLMAN, MONTAGUE, M.D.: Assistant Clinical Professor of Psychiatry, New York University College of Medicine, New York, New York.

WEBER, WILLIAM F., M.D.: Staff, Maimonides Hospital, Brooklyn, New York.

WEISMAN, MAXWELL N., M.D.: Director of Community Psychiatric Services, Department of Mental Hygiene, Baltimore, Maryland.

WOLBERG, ARLENE, M.S.S.: Postgraduate Center for Mental Health, New York, New York.

WOLITZKY, DAVID L., PH.D.: Consultant Psychologist, Hudson Guild Counseling Center, New York, New York.

YOLLES, STANLEY, F., M.D.: Director, National Institute of Mental Health, Bethesda, Maryland.

ZWERLING, ISRAEL, M.D., PH.D.: Associate Director of Psychiatry, Albert Einstein College of Medicine, Bronx, New York.

PREFACE

AS HAS NOW become customary, the Academy of Psychoanalysis has again taken the lead in coordinating its specialized field with two other disciplines of increasing interest and importance to the behavioral sciences: Communication (including Creativity), and its application to research, training and service in Community Psychiatry. This volume represents an integrated report of the three Scientific Sessions of the Academy devoted to these subjects in 1964: with the American Psychiatric Association in Los Angeles May 1-5, at its winter meeting in New York December 5-7, and with the American Association for the Advancement of Science in Montreal December 26-30.

The Publications Committee is once again deeply grateful for the erudition and expository skills of the contributors (the undersigned excepted) to this volume, and trusts that the rapidly expanding readership of this series will show a similar appreciation.

Jules H. Masserman, M.D.
Editor

Chicago, January 1, 1965

ACKNOWLEDGMENT

The quotations from Emily Dickinson used in this volume are from *The Poems of Emily Dickinson,* edited by Martha Dickinson Bianchi and Alfred Leete Hampson, and appear through the kind cooperation of the publisher, Little, Brown & Co., Boston, 1939, to whom thanks are extended.

Part I. Communication

Dynamics of Communication

ANXIETY: PROTEAN SOURCE OF COMMUNICATION

By JULES H. MASSERMAN, M.D.

IN THE BEGINNING was the Word—and that word, then as now, may well have been *Anxiety*. Since there is perhaps no more encompassing or versatile term in the English language, not excepting the almost equally poetic concepts of *narcissism,* or *love,* a survey of the multiple meanings of *anxiety* in their philologic, historical and clinical perspectives may well be our first task.

Aspects of Anxiety

Dictionary definitions, which reflect the semantic origins and evolutionary wealth of our language, are often more illuminating than technical strictures: thus, the *Unabridged Oxford* notes that anxiety was derived from the Latin *angere,* indicating an intuitive link between hostile impulses and fear of retribution. From this sprang three derivative meanings: *anguish,* (anticipated distress), *solicitude,* (empathy with a fellow-victim of fate), and *vital endeavor,* i.e., man's motivations to explore and control. But it is relevant to all three that German *Angst,* as used by Freud, also has the echoes of *limitation* and *constraint.*

Historically, the ancient Hebrews recognized the eternal struggle between *Yetzer Haráh* (Force of Destruction) and *Yetzer Hatóv* (Spirit of Good), just as the Hindus counterposed *Siva* and *Vishnu* and the Taoists, *Yin* and *Yang*. Without adequate reasons to be so, classical

Western sages were more confident both of man's intellectual powers and his moral self-determination: the Ionic philosophers anticipated modern physical and cosmic theories, Socrates taught that a "rational social order" would solve all of man's anxieties and Aristotle assured his complacent Academicians that man's knowledge of physical and biological phenomena would eventually establish his supremacy over nature. When this did not occur during the next eight centuries, St. Augustine, after he himself had tired of Manichaean mysteries and youthful Dionysian satisfactions, authoritatively advised everyone else to practice Ambrosian religious austerity. However, most men, when faced with environmental stresses they could not dissipate by currently available physical and technical means, characteristically reverted to what I have called our Ultimate (Ur) Defenses: i.e., dereistic denials, demographic deployments or deistic demonologies. In effect, men who could not cope with what they regarded as overwhelming threats got drunk and dissipated, or contracted opportunistic human alliances against other fellow-men, or sought magical power by cajoling or bribing a pantheon of gods conceived and controlled in man's image. A classical example of the first Ur-escape is furnished by Thucydides, who thus described the Plague of Athens in 480 B.C.

> "Men resolved to get out of life the pleasures which could be had speedily and would satisfy their lusts, regarding their bodies and their wealth alike as transitory. No fear of gods or law restrained them; for on the one hand, seeing that all men were perishing alike, they judged that piety and impiety came to the same thing; and on the other hand, no one expected that he would live to be called to account and pay the penalty for his misdeeds. On the contrary, they believed that the penalty already decreed against them and now hanging over their heads was a heavier one and that before it fell it was only reasonable to get some enjoyment out of life."

So also, when the Black Death had ravaged Europe for three hundred years despite man's futile devices of fires, philters and fasts, Michelangelo confessed "no thought is born in me that has not Death engraved on it," and George Wicher observed:

> "Some streets Tavernes had, rude revelles keeping.
> Some had churches full of people weeping.
> Within some houses, Psalmes and Hymnes were sung.
> With rayling and loud scoldings others rung."

Again, in a later outbreak, when all human life seemed to be as much in danger as it is in our present nuclear age, the three anxiety-relieving maneuvers of escape, aggression and mysticism were once more brought into full play: drunkenness was rampant; Queen Elizabeth, by divine right, hanged anyone who dared approach Windsor Castle; those less privileged killed Jews by the tens of thousands to please the gentle Jesus, and the revels of the Dancing Manias and the Black Sabbath spread throughout Europe. Only as man's trepidations abated, did gentler self-reassurance in the Classic manner reappear in the art of the Renaissance, and in the reawakened confidence of the *Nuova Scienza:* DaVinci now glorified bodily strength and beauty; Vives founded associationist psychology as the basis of all human knowledge and understanding and Nicholas Cusa and Nikolaj Kopernik founded modern astronomy—although, in a continuation of man's cosmic arrogance, they regarded the movements of the planets as also designed to predict the fate of the minor princelings who were their patrons. Only when the Religious Wars had almost completely abated did men resume the even more complacent egocentricities epitomized by Descartes' "*Cogito, ergo sum*" or Spinoza's more reflective solipsism:

> "I saw that all the things I feared, and which feared me had nothing good or bad in them save insofar as my mind is affected by them . . . . Fear cannot be without hope, nor hope without fear (for) all is Nature and God."

Leibnitz, in accordance with the new Baconian spirit, felt called upon to materialize this hope that all that exists is in the service of mankind by postulating that, whereas the ultimate "monads" which constitute the universe were "completely discrete and independent" they must nevertheless conform to the "laws of cosmic harmony" and thus create for humans, a "perfect social order." But Pascal, reverting to Jansenite mysticism, was once again less sanguine:

> "When I consider the brief span of my life, swallowed up in the eternity before and behind it, the small space that I fill, or even see, engulfed in the intimate immensity of spaces which I know not, and which know not me—I am afraid."

Regarding such premonitions of what is now called existential anxiety as being intolerable, subsequent emphathizers with man's dilemma again compounded more or less nebulous Ur-negations, Ur-norms or Ur-nepenthics: Berkeley denied reality; Kant restructured the universe according to his own convenient "categories"; Schopenhauer imperiously subjugated the world to his Will and Idea; Nietzsche extolled the amoral Superman; Marx advocated revolution leading to the abolition of all politico-economic controls, and Goethe (emulating Luther) used his Faustian inkwell to defy the Devil himself. With even greater monothetic fervor, Comte tried to reduce all sociology to a mathematical science (and ended by worshipping his dead wife as the Supreme Goddess of a new religion), Kafka pleaded piteously with man's unknown judges to specify their dark accusations; Santayana justified theology because of its "beauty"; Thoreau sought an earthly Walden, and Kierkegaard abandoned all reason in favor of an intensity of vital experience that was at once an acknowledgment and a defiance of a death-dealing God—an heroic stand that, minus even the comforts of a theistic faith, was also assumed variously by Sartre and the late Albert Camus. Yet despite such fervid philosophies man's manifold anxieties remained.

Psychoanalytic Concepts

Sigmund Freud, idolized by his family as a genius and raised in an Hebraic-Hasiddic tradition which included the notion of innate but unrecognized evil (the "Id"), preoccupation with sexuality ("libido"), the ritualized confessions to the rabbi as teacher ("emotional catharsis"), the hidden meanings of dreams ("symbolic interpretation"), and other proto-analytic concepts and practices (cf. D. Bakan), took readily to the introspective philosophies of his day and, in his own inimitable narcissistic-syncretic manner, colored them with his seventeenth-century faith in "pure reason" ("Where Id was, there shall Ego be"), a melange of Hartmann's arcane fantasies about the "unconscious", a transposition of the neural hierarchies of Hughlings Jackson (psychologized topologically as Id, Ego and Super-ego) and the quasierotic interpretations of Mesmeric phenomena he had observed in Charcot's and Bernheim's clinics. Then, with charismatically effective conviction, Freud also claimed that his new doctrines would modify many of the social and sexual prohibitions of his day, and thus imparted to his teachings (as Comte had done to his mystic version of sociology and as, with different pseudoscientific pretensions, Mary Baker Eddy had done with "Christian Science") not only the attractions of avant-garde social protest but also the trappings of a newly revealed and ritualized religion. In the latter respect, indeed, "classical" psychoanalysis offered a recorded and annotated Bible of Freudian gospels setting forth simple concepts of an intrapsychic Trinity, other dogmatic propositions exquisitely religious in form in that they were quite insusceptible of either proof or disproof (cf. Popper), a quasimystic ritual of indoctrination which held forth salvation through faith in the teachings of the Master, a roster of early Disciples (later to be individually canonized) and special studies of their exegetical writings, with the apocrypha of Jung, Adler, Rank, Sullivan, Horney and others fallen from grace specifically proscribed. Finally, after years of service as an acolyte, the seminary graduate receives a public anointment which confers on him the priestly privileges of hearing the confessional in a soundproof retreat sanctified by the order's ikons and images, receiving liberal tithes and comforting the laity by the absolution of sin and the revelation of ultimate truth. Parenthetically, the current movement among orthodox analysts (e.g., Gitelson, 1964) to extend these special privileges to our nonmedical confreres may in part be related to the likelihood that some (fortunately not all) members of these latter groups, in contrast to physicians with more disciplined scientific training and clinical experience, are more likely to believe that a solemn verbal incantation presented as an "analytic interpretation" can in itself change any cherished pattern of behavior. I have elsewhere defined what is usually called "insight" in such circumstances as that mutually happy state in which the patient shares the current illusions of his therapist. Of particular interest here, however, are Freud's own successive (but unfortunately less comprehensive) definitions of anxiety and its neurotic manifestations.

In 1894 Freud first attributed "anxiety neurosis" to a supposed "toxin . . . produced by frustraneous sexual excitement, the free-floating anxiety being ready to attach itself to any ideational content, such as phobias and obsessions." In later writings, Freud wisely abandoned such elementary biologizing and instead traced anxiety to the experience of external restraint and prohibition. "Anxiety as a signal of danger . . . mediated by the Ego" now *produced* repression rather than vice versa, the "impulses repressed" being those which might entail "loss of love, castration and social [internalized as Superego] condemnation." Freud then assured us that analytic therapy "will bring this unknown *instinctual* danger (again, not an acquired apprehension of external consequences) into consciousness and, in this way, the neurotic or instinctual anxiety becomes a *real* anxiety, anxiety over a *known* danger (italics mine). . . . Real anxiety—that is, fear of others, is then resolved in the transference which . . . should end in a heartfelt human relationship." However, since Freud also implied that all reality is "psychic" and equated all social punishment, including "loss of love," with symbolic "castration," these subdivisions of anxiety remained somewhat tenuous, as did his distinction between "real" and "neurotic" anxiety. The subsequent analytic literature on anxiety has unfortunately continued to be similarly equivocal.

Post-Freudian

Otto Rank's concept of birth trauma has often been interpreted too literally as physical suffering or injury to the neonate; actually, Rank meant what is now called "separation anxiety," whereas it was Freud who, in accordance with his first formula for anxiety, conceived of the birth trauma as a "flooding" of the newborn's immature nervous system with unaccustomed stimuli. Piaget and Odier, like Rank, thought of birth in terms of "abandonment" a precondition of what Fromm-Reichmann discerned in the adult as "loneliness."

For Alfred Adler and Karen Horney, anxiety represented the apprehension of a threat to, respectively, the "life style" or the "self-image" of the subject; as Erik Erikson has more recently pointed out, this could lead to an agonizing confusion of "partial identities" in adolescents who have not been helped to develop clear values and goals.

Other current combinations and permutations of these concepts may be summarized as follows:

Roy Grinker, Sr. believes that whereas "animals may become alert and vigilant, only self-reflective humans can know anxiety as conscious dread of the future"—though this would in turn raise the question as to what else the animals could be "vigilant" about. Paul Tillich is predictably more somberly reflective: "Normal anxiety is equatable with the finiteness of man," whereas for Rollo May "Anxiety is occasioned by a threat to *any* value considered essential"—hence, parenthetically, anxiety-relieving drugs may prevent or confuse effective conduct.

Kurt Goldstein measures "fear, which leads to adaptation" against "anxiety, which may result in catastrophic paralysis," hence psychotherapy should be deliberately designed to convert vitiating anxieties into realistic fears leading to appropriate action. Yet on the basis of extensive military experience, A. J. Glass has once again come full circle in questioning the validity of sharp distinctions between fear and anxiety, since both cover widely overlapping ranges of helplessness in the face of variably sensed insecurities, both are inextricably mixed with trepidation, dread, anger or other equally protean "affects," both are expressed in identical physiological reactions, and both are relieved when the anxious or fearful subject believes, with whatever illusions, that his safety and well-being have been restored through the efficacy of his Ur-defensive gadgets, groups or gods.

Ur-Anxieties

If, then, the *essential* features of all of these concepts were to be combined, anxiety could be operationally defined as a repressed apprehension of earlier dangers easily re-evoked by situational symbols to which the individual has become psychologically hypersensitive and somatically hyperactive. Over a decade ago (1953), I attempted to correlate this definition with the deeper and perhaps ultimate Ur-sources of man's trepidations as follows:

> "Man's Ur-anxieties are three: first, his abhorrence of physical injury and death; second, his uncertainty as to the reliability of his human alliances; third, his utter rejection of the thought that perhaps he is, after all, little more than a cosmic* triviality. It is equally significant that these triple trepidations of man also motivate his principal modes of presumed mastery: first, his attempt to subjugate his material milieu through various sciences and technologies; next, his efforts to safeguard his social security by familial, economic, political and military alliances; and finally, his endeavors to encompass the entire universe in his philosophic and religious systems."†

Physiologic Substrate of Anxiety

Since, then, anxiety has connoted almost every conceivable combination of motives, experience,‡ affects and ideas, it is hardly surprising that the more carefully conceived and controlled the search for

---

*My secretary, perhaps more appropriately, first typed this as "comic."

†T. R. Sarbin has very recently pointed out that the term anxiety was not used in psychology before Freud's writings and that, like Descartes' concept of "soul," it is still merely a metaphorical reification unnecessarily postulated between distal events [in a "real" world?] and their physiological, subjective and behavioral consequences in a designated individual. Sarbin, without reference to my earlier concepts of the three basic Ur-defenses, expands the sources of anxiety to five "cognitive strains" or threats; i.e., to self-concept (*what* am I?) time space orientations (*where* am I?) social function (*who* am I?) normative performance (*how well* am I doing?) and transcendental concern (*why* am I?)

‡The observations of Divenberg and Whembey throw an interesting sidelight on the "familial" transmission on early experiences: when rat pups are raised by adult mothers who had been mishandled in their own infancy, the pups show various behavioral disturbances not shown by controls raised by normal natural or foster mothers.

its anatomical nidus or physiological accompaniments, the less definitive the results. In this regard history again furnishes the proper perspective. Two and a half millennia ago, Aristotle traced the source or "entelechy" of all behavior to the heart, Erasistratos to the cerebellum and Herophilus to the third ventricle; more wisely, Democritus distributed the functions of the human soul among the liver (desire), heart (anger), and brain (reason). Galen taught that "will and feeling" originated in the rhombencephalon, whereas Descartes, always seeking concentric symmetry, designated the pineal. Turning to the periphery, William James (1884) proposed that all "emotions," including anxiety, were perceived as secondary to visceral reactions;* almost simultaneously, G. C. Lange (ergo, the James-Lange theory) stressed the vascular component, and a decade later John Dewey and subsequently H. Kantor and J. Drever called attention to the necessity of bodily *frustrations* in the genesis of affect. However, all such visceropoetic speculations were challenged by the clinical observations of Sherrington (1900) and the experimental work of Cannon, Lewis and Britton (1927) which demonstrated that interference with the afferent somesthetic or efferent spinal pathways to the viscera did not markedly influence emotional reactions; on this and other evidence, H. Head (1921) returned to the thalamus as the affective center, and W. Cannon (1931) and P. Bard added the concept of emotional-ideational resonance *via* thalamo-cortical radiations. Leeper (1948) pointed out that since "affects mobilized action" rather than disorganized it, distinct cerebral pathways for the various "emotions" were unlikely; nevertheless, Arnold (1950) assigned separate cortical routes for fear, anger and sexual excitement.†

Diencephalon

Meanwhile, on the basis of observations by Cannon, Bard, Rioch, Ranson et al. that stimulation of the hypothalamus produced "sham rage" or "pseudo-anxiety" in cats, this portion of the diencephalon had for a time again been designated by Alvarez, Serota and others as the "center of emotion." However, stimulation of the hypothalamus

*Again *e-moto*, to move out.

†For a detailed and documented review with a somewhat different focus, see the excellent chapter on Emotion by D. B. Lindsley in the *Handbook of Experimental Psychology*, S. S. Stevens, ed. New York, Wiley and Sons, 1951, p. 473ff. In contrast, many psychoanalytic writings on anxiety are like a child's game of jacks: one rounded, malleable concept is bounced, and many other knobby ones are swept together in a conglomerate mass before being once again scattered.

in human subjects by J. C. White produced only cardiac slowing and drowsiness, and a series of animal studies by Masserman et al. indicated that whereas the hypothalamus was undoubtedly a coordinating way-station on the viscero-motor pathways for emotional expression, it was in all probability not a "center" that either "originated" or "experienced" subjective affect.

Limbic Circuits

As long ago as 1878, Broca had noted that the limbic lobes were prominent in all mammals capable of versatile "emotional" responses, and von Economo, a generation before Moruzzi and Magoun, had attributed activating functions to the peri-aqueductal gray and mesencephalic reticulum. In 1937, J. W. Papez suggested that the rhinencephalon, in which the neural circuits of olfaction had been primarily concerned with escaping from danger, securing food and finding mates, now also mediated "emotions" via connections to the hippocampus, amygdalae, cingulum, cuneus and diencephalon. P. McLean (1955) then proposed that these circuits constituted a "visceral brain" which receives affectively charged stimuli significant to sex and survival through the brain stem and midthalamic nuclei, regulates their intensity in the reticular system, marshals appropriate visceral responses *via* the fornix and hypothalamus, enlists the endocrine organs through tubero-pituitary tracts and hormones, and then relays these neural transactions to the cortex for conscious ideational evaluation. Striking illustration of the interrelationships of brain function and experience appeared in our own work (1950-60), which showed that lesions of the dorsomedial thalamic nucleus released aggression to a far greater degree in adult animals that had been exposed to neurotigenic conflicts than in normal controls, whereas bilateral amygdalectomy produced an irreversible Klüver-Bucy syndrome only in young animals. In other investigations, J. Olds, K. Pribram, J. Delgado and others isolated special regions in the septum and lateral amygdalae which, when stimulated, elicit evidences of pleasure and repetition seeking as contrasted with other points in the tegmentum, fornix and posterior hippocampus which induce avoidance; however, these neural nodes too, change or even reverse their functions under overstimulation, external satiation and various other experimental contingencies (N. Miller).*

---

*For additional discussion and bibliography, cf. Masserman, J. H., *Principles of Dynamic Psychiatry*. Philadelphia, W. B. Saunders, 1962.

Experiential Effects

But perhaps the most clinically significant animal studies were behavioral rather than strictly neurophysiological. Begun by Pavlov and continued by Shenger-Krestovnikova, Gantt, Liddell, Jacobson, Hoff, Shenker, myself et al., these showed that, irrespective of lesions or stimulations of the hypothalamus, thalamus, cortex or other portions of the central nervous system, most of the manifestations of "anxiety" and its accompanying behavioral deviations (e.g., symbolic inhibitions, phobic and compulsive aversions, somatic dysfunctions and disturbances in social relationships), could be predictably induced by subjecting an animal to disruptive conflicts between incompatible motivations and opposed patterns of adaptation.*

As examples of recent work, Curt Richter observed that hungry rats offered poisoned food froze into bizarre postures or ran to exhaustion if a wheel was available; at "higher" libidinal levels, A. Wikler reported that pups subjected to an electric shock on their first attempt at coitus developed a "sexual neurosis." i.e., life-long deviations in sexual patterns. It is relevant here that disturbed maternal and social avidities in adulthood could also be shown to result from infantile deprivations; Green and Gordon summarize their observations in this regard as follows:

> Visual exploration was studied in maternally reared and maternally deprived monkeys. When an animal pressed a bar an opaque screen was raised providing a brief view of either of a pair of stimuli. Subjects reared by their mothers pressed more to see animate than inanimate objects. With increasing age, the number of bar-pressing responses decreased for an adult female stimulus, increased for an age peer and for food, and remained low for geometric forms and an empty chamber. *Maternally deprived subjects established uniformily low response levels to all stimuli.*

Electroencephalography

R. Cohn has observed that anxiety may abolish alpha rhythms or induce high-voltage beta waves; in animals, these affects may be mediated through the ascending reticular formation (D. B. Lindsley, J. Bowden and H. W. Magoun). C. E. Wells and H. G. Wolff noted that the abolition of alpha-waves by an auditory signal previously conditioned to a light stimulus was diminished both in patients with

*Our experiments have shown that such conflicts need *not* involve fear; "positive" but mutually exclusive goal-seeking patterns may be almost as neurotigenic.

brain damage and in those with prolonged and severe anxiety, and inferred that anxiety may thus produce organic cerebral defects.

Biochemical Substrates

The most reliable as well as the most obvious physiological indices are the rapid heart rate, raised pulse and systolic blood pressure, vasomotor instability, horripilation and skeletal tensions that accompany the "anxiety syndrome" described in the clinical sections below. As to their biochemical substrate, Grinker and his associates found that an increased excretion of hippuric acid and of 17 keto-steroids were the *only* predictable variables in severe emotional states,* whereas hyperglycemia during "emergency reactions" occurred only in diabetics (I. A. Mirsky). The contention that adrenin secretion is increased when "anger is turned inward as anxiety" whereas nor-adrenin increases when "anger is expressed outwardly in aggression" (D. Funkenstein et al.) has been called into question on both psychological and physiological grounds. However, an excess of acetylcholine as indicated by an increased galvanic skin response (GSR)† may occur in "mild anxiety" (as it does also in deep concentration, sexual excitement, etc.), in contrast with peripheral vasoconstriction in "white panic."

Disturbed Feedbacks

B. S. Lee, K. U. Smith, W. M. Smith, my associates and I, and others, using special sound recording and video apparatus, have demonstrated that a delay of only a fraction of a second in the auditory or visual feedback of a human subject's own voice or motion may produce sensorial confusion, affective apprehension and marked and persistent disturbances in performance. On the basis of such data, Bowers et al. concluded that any serious disturbances in accustomed configurations of sensation, arrangements, sequences, relationships or systems of self-and-reality produce insecurity and hence anxiety—an inference generalized to the field of interpersonal communication by Jurgen Reusch and J. G. Miller (*v.i.*).

*Gibbons found a similar correlation between the intensity of clinical depression and the cortisol content of plasma and urine; so also, Kurland noted an absolute rise in 17-ketogenicsteroids and a relative increase in urinary 17-ketosteroids and 17-hydroxycorticosteroids, especially in the morning hours of diurnal melancholia.

†Mordkoff reported that subjects exposed to an emotionally stimulating film of Australian puberty rites showed greatest alterations in skin resistance, less in pulse rate, and least in respiration.

Clinical Aspects of Anxiety

Nosology

The current Standard Nomenclature of the American Psychiatric Association attempts to telescope all of Freud's successive notions of anxiety (*v.s.*) into this somewhat tortured definition:

> "Anxiety in psychoneurotic disorders is a danger signal felt and perceived by the conscious portion of the personality. It is produced by a threat from within the personality (e.g., by supercharged repressed emotions, including such aggressive impulses as hostility and resentment) with or without stimulation from such external situations as (sic) loss of life, loss of prestige or threat of injury."

The latest revision of the American Psychiatric Association Diagnostic Manual retains only one nosologic entity: *Anxiety Reaction,* characterized as "diffuse and not related to situations or objects as in the case of phobic reactions [and] to be differentiated from normal apprehensiveness or fear." In clinical practice, however, the following variants, listed in previous editions, are still frequently used:

Anxiety Syndrome: Characterized by severe, unformulated apprehension and its physiological concommitants: generally cardiac palpitation, irregular or rapid respiration (which may lead to hyperventilation and secondary hypotension with syncope), "globus hystericus" (pharyngeal constriction), trembling, flushing, sweating and splanchnic spasms ("flutterings in the stomach") that may require repeated pharmacological control with ataractic or sedative drugs.

Anxiety Hysteria: When accompanied by marked sensorimotor ("conversion"), functional amnestic, or persistent organic (psychosomatic) dysfunctions which may necessitate ancillary medical or surgical intervention: e.g., as in gastric ulcer.

Anxiety Neurosis: Recurrent episodes as above, with phobias, obsessions, compulsions and other more deeply ingrained "neurotic" traits and deviations of character amenable, if at all, only to individualized long-term treatment.

Therapy

Anxiety has one final therapeutic connotation: a state of tension, ranging from unpleasant to unbearable, which induces a patient to seek professional aid; indeed, Freud insisted that analysis "cannot take place in the *absence* of anxiety." But since the myriad somatic and behavioral concommitants of anxiety as reviewed above add a multitude of biological, personal, metapsychological and social dimensions to an already almost incredibly complex field of human discourse, it becomes evident that no limited or stereotyped mode of "treating anxiety," although simple prescriptions are always tempting to the bewildered, can cover even a portion of the intricacies and contingencies involved. In broader statement, the alleviation of "anxiety," (or its elicitation in the sense of inducing optimal modes of alertness, concern and effectiveness of response) is integral with all psychiatric therapy, and may most briefly be stated as follows:

In psychiatry, therapy comprises the most efficient utilization of *every* pharmacologic or other medical means* to relieve the patient's symptoms, and *every* ethical mode of communication and influence—including subjective analysis, reorientative dyadic transactions and guided resocializing experiences†—to help him realize that his previously "neurotic" patterns of behavior, although perhaps once suitable, are no longer either necessary or advantageous, whereas more effective, creative and culturally adaptive modes of conduct will be found to be on the whole more pleasurable and profitable—and in no sense a stultified abandonment of individuality. Our techniques of therapy (Gk. *therapeien* = service) will thus attempt insofar as possible to meet the patient's fundamental seekings (Ur-faiths) as reviewed above: first, through the restoration of his bodily strengths and skills; second, the recultivation of his human companionships, and third, the reinvocation of his trancendent philosophical-theological securities. Can any mortal do more?

REFERENCES

Adler, A.: The Practice and Theory of Individual Psychology. New York, Harcourt Brace and Co., 1924.

---

*In this connection, however, evidence is accumulating (cf. Brill et al.) that meprobamate and the phenothiazines add little or nothing to the effects of psychotherapeutic sessions, whatever their frequency and duration, in relieving anxiety and its symptomatic expressions.

†The latter therapeutic factor has recently been particularly well stressed by Jay Haley, who points out that, despite our "metarule" of permissiveness, in actual practice we impose many financial, interpersonal and other powerful sanctions on the patient which induce or compel him to improve his individual and social behavior.

BAKAN, D.: Freud and the Hassidic Tradition. Chicago University of Chicago Press, 1962.

BARD, L.: Neural mechanisms in emotional and sexual behavior. Psychosom. Med. 4:171, 1942.

BASOWITZ, L. ET AL.: Anxiety and Stress. New York, McGraw-Hill, Co., 1955.

BRILL, N. Q.: Controlled study of psychiatric out-patient treatment. Arch. Gen. Psychiat. 10:581, 1964.

CANNON, W.: Again the James-Lange theory of emotion. Psychol. Rev. 38:281, 1931.

COHN, R.: The influence of emotion on the human EEG. J. Nerv. & Ment. Dis. 104:351, 1946.

DEVENBERG, V. H., AND WHIMBEY, A. E.: Behavior of adult rats as modified by the experiences their mothers had as infants. Science. 142:1192, 1963.

ERIKSON, E. H.: The problem of ego identity. J. Am. Psychoanalyt. A. 4:56, 1956.

FAIRBANKS, G., AND GUTTMAN, N.: Auditory feedbacks. J. Speech & Hearing Res. 1:12, 1958.

FESTINGER, L.: A Theory of Cognitive Dissonance. Palo Alto, Stanford University Press, 1959.

FREUD, S.: The Problem of Anxiety. New York, Norton and Co., 1936.

——: Collected Papers, Volume 1. New York, Basic Books, 1959.

FROMM-REICHMANN, F.: Loneliness. Psychiatry 22:1-15, 1959.

GIBBONS, J. L.: Cortisol secretion in depressive illness. Arch. Gen. Psychiat. 133:572, 1964.

GLASS, A.J.: Psychotherapy in the combat zone. Am. J. Psychiat. 110:725, 1959.

GLESER, G. C. ET AL.: Anxiety scale applicable to verbal samples. Arch. Gen. Psychiat. 5:593, 1961.

GREEN, P. C., AND GORDON, M.: Maternal deprivation: Its influence on visual exploration in infant monkeys. Science 145:292, 1964.

GREENACRE, PHYLLIS: The predisposition to anxiety. Psychoanalyt. Quart. 10:667, 1941.

GRINKER, R. R., SR.: Anxiety as a significant variable. Arch Gen. Psychiat. 1:537, 1959.

GROTJAHN, M.: Freud's Letters to Oskar Pfister. Personal Communication, 1927.

HALEY, J.: Strategies of Psychotherapy. New York and London, Grune & Stratton, 1963.

HEAD, H.: Release of function of the nervous system. Proc. Roy. Soc. London 92:184, 1921.

HORNEY, KAREN: The Neurotic Personality of our Time. New York, W. W. Norton & Co., 1937.

KIERKEGAARD, S.: The Concept of God. Princeton, Princeton Unviersity, 1944.

KURLAND, H. D.: Steroid excretion in depressive disorders. Arch. Gen. Psychiat. 10:554, 1964.

LANGER, W. L.: The black death. Scient. Am. 210:115, 1962.

LEE, B. S.: Delayed auditory feedbacks. J. Acoust. Soc. Am. 22:824, 1950.

LILLY, J. C.: Distress call of the bottlenose dolphin. Science 139, 116, 1936.

LINDSLEY, D. B., BOWDEN, J., AND MAGOUN, H. W.: Effect upon the EEG of acute injury to the brain stem activating system. EEG Clinical Neuro-physiol. 1:975, 1944.

Magoun, H. W.: An ascending reticular activating system in the brain stem. Arch. Neurol. & Psychiat. 67:145, 1952.

Masserman, J. H.: Faith and delusion in psychotherapy. Am. J. Psychiat. 110:324-333, 1953.

——: Practice of Dynamic Psychiatry. Philadelphia, W. B. Saunders, 1955.

——:Ethology, comparative biodynamics and psychoanalytic research. *In* Science and Psychoanalysis, vol. III, J. H. Masserman, ed. New York and London, Grune & Stratton, 1960, pp. 20-80.

——: The office therapy of psychosomatic disorders. Arch. Gen. Psychiat. 3:320, 1960.

——:Principles of Dynamic Psychiatry, 2nd ed. Philadelphia, W. B. Saunders, 1962.

——: A tachistoscopic glance at recent advances in the neurophysiology of behavior. Israel Ann. Psychol. 1:244, 1963.

——: Contribution of experimental psychiatry to the art of healing. Psychoanalyt. Rev. 50:103, 1953-1964.

——: Behavior and Neuroses, 2nd ed. New York, Hafner, 1964.

McLean, P.: A proposed neural mechanism of emotion. Arch. of Neurol. & Psychiat. 38:725, 1937.

Mordlsoff, A. M.: The relationship between psychological and physiological response to stress. Psychosom. Med. 26:135, 1964.

Odier, C.: Anxiety and Magic Thinking. New York, International Universities Press, 1955.

Pascal, B.: Thoughts. New York, Peter Pauper Press, 1946.

Persky, I. et.al.: Adrenal cortical functions in anxious human subjects. Arch. Neurol. & Psychiat. 76:549, 1956.

Popper, K.: The logic of scientific discovery, New York, Philosophical Library, 1958.

Powers, W. T., Clark, R. K., and McFarland, R. L.: A general feedback theory of human behavior. Precept & Motor Skills 11:309, 1960.

Ruesch, J.: Disturbed Communication. New York, W. W. Norton and Co., 1957.

Sarbin, T. R.: Anxiety: Reification of a metaphor. Arch. Gen. Psychiat. 10:630, 1964.

Smith, K. U., and Bloom, R. J.: Appl. Psychol. 40:302, 1956.

Smith, W. M. et. al.: Delayed visual feedbacks and behavior. Sciences. 1330:1013. Also, Scient. Am. 204:86.

Spinoza, B.: Ethics. London, Everyman, 1910.

Tanger, W. L.: The Black Death. Scient. Am. 210:115, 1963.

White, J. C.: Automatic discharge from stimulation of the hypothalamus in man. Research Publication for Association of Research in Nervous and Mental Disease. 20:854, 1940.

CONTRIBUTIONS TO COGNITION FROM PSYCHOANALYTIC THEORY

By SILVANO ARIETI, M.D.

COGNITION is or has been, up to now, the Cinderella of psychoanalysis and psychiatry. No other field of the psyche has been so consistently neglected by clinicians and theoreticians alike. Isolated studies and manifestations of interest have not so far developed into a definite trend. And now, that some sustained interest in cognition seems finally to emerge from decades of slumber, a survey of past and recent contributions finds a body of material difficult to organize and even more difficult to assimilate.

Freud and the Freudian School

Most authors agree in considering the discovery and the study of the primary process Freud's fundamental contribution to the field of cognition. The description of the primary process and its separation from the secondary process, as illustrated in Chapter 7 of the book *The Interpretation of Dreams,* remain classic.[12]

Ernest Jones wrote:

> ". . . Freud's revolutionary contribution to psychology was not so much his demonstrating the existence of an unconscious, and perhaps not even his exploration of its content, as his proposition that there are two fundamentally different kinds of mental processes, which he termed primary and secondary. . ."[17]

Freud gave the first description of the two processes and tried to differentiate the particular laws or principles which rule the primary process only. He called the primary process primary because, according to him, it occurs earlier in the ontogenetic development, and not because it is more important than the secondary. As a matter of fact, Freud stressed the point that the secondary process becomes the dominant one in the normal adult.

Freud elucidated very well two mechanisms by which the primary process operates. They are the mechanisms of displacement and condensation. They are now of such common knowledge, that we tend to forget that they were unknown prior to Freud and that a great creative ability was necessary to discover and differentiate them.

The founder of psychoanalysis illustrated these two mechanisms especially in the symptomatology of the psychoneuroses and in the symbolism of dreams.[11,12]

In his first major theory, generally called the topographic theory, Freud considered the primary process as a way of functioning of the unconscious, as well as of mental life in early childhood, before the system preconscious comes into being.

In his second major theory, generally called the structural theory, Freud did not limit the activities of the primary process to the id, but extended them to the ego and the superego. In other words, according to Freud, each of the three divisions of the psyche can operate according to either of the two processes.

This great breakthrough represented by the discovery and first description of the primary process, did not lead to a series of cognitive discoveries by Freud or by the Freudian school. The arrest of progress in this field is, in my opinion, to be attributed to several factors. Freud did not maintain a great interest in the primary process as a mode of cognition, but only as a carrier of unconscious motivation. Unconscious motivation is probably the most important aspect of psychoanalysis, and the fact that Freud should be so preoccupied with it is fully understandable. However, inasmuch as motivational theory, in the Freudian system, came to be interpreted in the function of the libido theory, the primary process came to be studied not in a framework of cognition, but in a framework of energetics.

Primary and secondary processes came to be considered primarily not as two different ways of thinking, but as two different ways of dealing with cathexes. In the primary process the cathexis is described as free. In other words, cathectic quantities of energy are easily shifted from some objects to others. Inasmuch as this shifting may easily occur from realistic and appropriate objects to unrealistic and inappropriate, the primary process becomes an irrational mode of functioning.

These points of view leave many unexplored aspects, especially those which pertain more closely to cognition. What is ignored in these studies is that a shifting of libido requires the existence of alternative cognitive elements. The psyche must have available the mental representations of various objects, to which the cathexis is shifted. Furthermore, when the libido is shifted to another object, this shifting or displacement is not due to chance, but to a cognitive relation between the original object and the substitutes. For instance, if in dreams the cathexis is shifted from the vagina to a box, this shifting is not casual, but due to a special type of cognition, which first is able to represent mentally a vagina and a box, and secondly sees or intuits a special relation between the function and shape of the vagina and of the box, so that the latter can replace the former. In Freudian theory the displacement is seen almost exclusively as a shifting of mental energy. It is not attributed to a cognitive creation of mental alternatives and to the immediate grasping of a relation between these alternatives.

Moreover, this shifting of mental energy may be viewed as the function of the mechanism of attention or lack of attention. Whereas Freud in his early writings paid some attention to the phenomenon of attention, he later ignored it. In a state of sleepiness, drunkenness, or psychosis, the span of attention is reduced and distractibility toward emerging collateral ideas is increased. For instance, people who try to say prayers while they are awakening or falling asleep, have difficulties in completing the required sentences, as emerging distracting thoughts interfere.

If the study of the primary process did not lead Freud and the Freudians to additional discoveries in the field of cognition, the study of the secondary process was even less fruitful. Gill,[13] Arlow and Brenner,[7] to mention just a few, have systematized the findings of Freud but have not added new prospectives. As far as the secondary process is concerned, it is often repeated that it is characterized by fixed or bound cathexis or by the fact that there is a structuralized delay between the drive or derivative drive and thought.

Perhaps the greatest systematizer of Freud's theories is David Rapaport,[21,22,23,24] who has incessantly tried to reconcile the diverging points of view, and to fill the gaps in the consistency of the system. Rapaport describes Freud's primary model of thought as consisting of the following steps: mounting drive-tension, absence of drive-object, hallucinatory image of it.

Because of the delay in the gratification discharge, the memory-traces of the gratification experience are cathected to perceptual intensity. In the words of Rapaport "the delay of gratification-discharge by motor-action becomes the cradle of 'conscious experience.' " Memories are representations of drives and become organized around the drives which they represent. They become thus drive-representations and are indicators of mounting drive-tension. At this early stage the primary psychic structure is characterized by the mobility of cathexis and by the mechanisms of displacement, condensation and substitution. According to Rapaport, these representations or rather drive-cathexes will then follow two major patterns: they will either become repressed or will change to a bound form.

In primal repression the drive and its cognitive representation are denied access to motility and consciousness. The repression is determined by an energy charge used against the drive-cathexis. This energy charge is the countercathexis. The system of the countercathexis is conceptualized as the *censorship.* What is the origin of the energy of the countercathexis? It is the energy of the drive-cathexis itself, which is denied emergency to consciousness. In other words, the energy which was supposed to accompany the primal drive-representation to its discharge is, according to Freud-Rapaport theory, now used to repress the primal drive-representation.

Rapaport states that the other fate of drive-cathexes, that is, their transformation to a bound state, is far less understood. In other words, the more organized and the more close to the secondary process a psychic construct is the less it is understood by psychoanalytic theory. In the secondary process the drive does not reach discharge by a short-cut route, nor is it fully repressed, but reaches its destination by a detour. Now memory traces are not used because they belong to the same drive, but rather because of "*a connectedness along the pathways in reality* toward the need-satisfying object."[22] The connections begin "to take on the form of reality connections: space, time, continuity, similarity." Rapaport, however, does not study this "connectedness" along the pathways of reality, or the elaborations or abstract derivatives of the concretely connected elements.

Rapaport attempts to study conceptual thinking, attention and anticipation, but again as a function of the vicissitudes of the cathexis. In his studies of memory, however, he makes the pertinent observation that memory cannot be considered exclusively an autonomous function. As a matter of fact, manifestations of the drive-organization of memory —such as slips of tongue, dream-work, forgetting due to repression— reveal the drive organization of memory.

Rapaport's extreme loyalty to Freudian conceptualization restricts the breadth of his thinking. Although in his remarkable book *Organiza-*

tion and Pathology of Thought he presents a few pre-Freudian and non-Freudian approaches, he selects only those which can be integrated or reconciled with the Freudian theory. He ignores completely Giambattisto, Vico, and Susanne Langer and makes only a passing reference to Ernst Cassirer. Although he seems to have a great deal of respect for the developmental psychologist Heinz Werner, he quotes him sporadically and only in support of Freudian theory and ignores completely Werner's fundamental ideas. Summarizing, Rapaport too is basically concerned with the cathectic charges of the cognitive constructs, not with cognitive or symbolic forms.

Although the secondary process, like the primary, is supposed to occur in all the three divisions of the psyche, it is especially in the ego that one would expect its clearest manifestations. One would thus anticipate that the greatest contributions to the knowledge of the secondary process would be the work of the ego psychologists. Here again we are bound to be disappointed. Hartmann, Kris and Loewenstein[15,16] had to do something which seems unnecessary in the field of psychology in general, but which, on account of Freud's emphasis, was necessary in the field of psychoanalysis: namely, to re-establish and reaffirm the importance of ego structure, secondary process thinking, perception, etc. Hartmann[14] also asserted that the ego is not a derivative of the id, but has autonomous origin and functioning.

For the purposes of this paper, it is not necessary to review the contributions of the ego psychologists, because again their main concern was with the origin and distribution of the cathexis necessary for the functioning of the ego, not with the cognitive forms and mechanisms of the ego. Energetics is for them almost equated to psychodynamics. In their numerous studies of the ego and of the secondary process they could not duplicate the success of Freud's study of the primary process.

Neo-Freudian Contributions

One would be inclined to think that the neo-Freudian schools would have paid more attention to cognition, inasmuch as in their theoretical framework they do not stress the primitive or instinctual aspects of the psyche. And yet, we look in vain for major contributions to cognition even by such neo-Freudian leaders as Sullivan, Horney, Fromm, etc. These innovators concentrated on the importance of social factors in man's development and motivation and paid little attention to cognition.

Occasional cognitive studies have been carried out by several people of different neo-Freudian orientations, for instance, by Harold Kelman of the Horney school. However, it is especially at the William Alanson White Institute of New York that we find a deep interest in cognition.

We could divide the investigators there into three groups: (1) Those who have worked predominantly on prelogical experience; this group includes Tauber and Green. (2) Those who have worked predominantly and in different ways in the field of communication. This group includes Edwin Weinstein, Rose Spiegel and Joseph Jaffe, the latter by empirical and not a clinical-theoretical approach. (3) The third group consists of authors who have studied cognition predominantly from a developmental point of view; this includes Ernest Schachtel and Silvano Arieti.

Tauber and Green[31] have studied cognition, insofar as it concerns what they have called prelogical experience. According to them prelogical thinking is "part of the basic endowment of man throughout life, which operates continuously." It operates not only in dreams, but also in creativity, in fringe phenomena like hunches, beliefs and intuition, in subthreshhold phenomena, in altered states of consciousness, in extrasensory perception, and in the transference and countertransference situation. Tauber and Green do not make use of Freud's concept of primary process, nor do they describe the different forms of prelogical cognition or the laws by which these forms of experiences are regulated. However they do affirm, and in my opinion rightly so, that early perception in the baby is characterized by fragmented, dreamy elements hinting at connectedness with the outside world. Rather than on the study of the forms of cognition, these authors have focused on the interpersonal aspect of prelogical experience. This interpersonal approach to what seems intrapsychic gives originality to their contributions.

In his studies on language and communication, Edwin Weinstein[35,36] has relied on his neurological as well as psychiatric-psychoanalytic clinical experiences. Weinstein also makes larger use of Sapir's classification of symbolism into referential and experiential types. Referential symbols are arbitrary designations coined by the community for purposes of communication and reference. Typical of this category are numbers, letters, Morse code, etc. Experiential symbols are those which express predominantly feelings and emotions. In situations of stress, as well as in psychiatric illness, communication becomes more and more referential in character.

Weinstein has studied communication especially in patients suffering from organic conditions of the central nervous system, and has shown how the changes in communication are more indicative of a change to an experiential type of symbolism than of a real defect. In such conditions as anosognosia, or denial of illness, and amnesia, the need to avoid the catastrophe which would be brought about by the knowledge of one's condition is so overpowering as to elicit amazing denials. Using the expressions of three other authors, whose contributions he has admirably integrated, we may say that Weinstein's theory is the following: in order to avoid what Kurt Goldstein called a catastrophic reaction, the patient achieves what Freud called wishfulfilling denial by adopting what Sapir called the experiential type of communication.

According to Rose Spiegel[30] the communication circuit consists of the sender of the message, the message, the receiver of the message and his response to the message. "The message . . . is an extension of the sender, and how it is received is a measure of the recipient."

Rose Spiegel divided faulty interpersonal communication into five categories: (1) grossly destructive communication, characterized by hostile intent, which may be conscious or unconscious; (2) authoritarian communication, whose most blatant examples are the "brain washings" of the police-state technique, but whose most common forms occur in everyday life. For instance, when a parent says "you are to talk when I tell you," he or she uses authoritarian communication; (3) disjunctive communication, whose predominant purpose is "confusion, disruption or derisiveness"; (4) pseudocommunication, as in double talk; (5) noncommunication, which occurs in situations which are communicative only in appearance.

Spiegel studied in detail the alterations of communication in various psychiatric conditions. In hysteria there is a resymbolization of an intolerable emotional experience into other terms: for instance, in conversion hysteria there is a resymbolization into somatization; in anxiety hysteria, into phobic resymbolization of anxiety; in hysterical alterations of consciousness, into resymbolization of one's sense of self; in hysterical character, into resymbolized acting out in the form of a different personality. Spiegel's accurate and detailed interpretations of communication in schizophrenia and depression cannot be detailed here.[20,30]

As I have indicated, Schachtel follows a developmental approach. As a matter of fact, the title of his major work, "Metamorphosis" refers to the developmental transformations which occur from birth to maturity.[27] Only a part of this book deals with cognition. According to Schachtel, perception occurs in two basic modes. One mode is that of autocentric perception, in which the perceiver centers on what happens to himself and focuses on the sensory experience. The other mode is the allocentric in which the perceiver focuses on the object as a separate entity. For Schachtel perception is motivated not only by the need to reduce the tension of a drive, but by an actual interest in the outside world.

Two particular topics in the field of cognition have been intensely studied by Schachtel. One of them is focal attention—the capacity to center one's full attention on an object so that the object can be perceived and understood from many sides. Focal attention permits the emergence of the object world, that is, of reality. It is a characteristic of what in the Freudian system is called the secondary process. Thus the inability to focus on an object from many angles, which Schachtel describes in the child prior to the acquisition of focal attention, may be compared to the shifting of cathexis, described by the Freudian school in the primary process. However, Schachtel's mechanisms of focal attention or of lack of this focal attention seem to apply only to conscious phenomena. Schachtel strongly rejects Freud's idea that thinking is a substitution for the hallucinatory wish. He believes that thought has "two ancestors": one is the motivating need, the other is the capacity for object interest.

The other topic of cognition which Schachtel has intensely studied is the one in which he probably has made the most important original contributions: childhood amnesia.[26] Schachtel believes that childhood amnesia is not caused by repression or by some kind of censorship, but by the fact that the cognitive schemata of adult life are not available to the child, so that he cannot preserve a memory of his early experiences. Even if these schemata were available, they would not be able to reproduce the quality and the intensity typical of the experiences of early childhood.

Schachtel's important contributions have reaffirmed that the main purpose of man's psychological life is not merely reduction of need—a reductionist point of view of the Freudian school—but also an interest for the world of reality, an openness to the world. Without perhaps being aware of it, Schachtel in all his writings seems to be under the influence of the German philosopher, Max Scheler, and of the latter's conception of openness to the world.

Although Schachtel's contributions to cognition are important and original they have some limitations. His developmental approach does full justice only to the ontogenetic development. He pays little attention to the phylogenetic and no attention whatsoever to what Werner called the microgenetic development (*v.i.*). His approach is thus similar to Piaget's, except that Piaget studies the highest levels of cognition, whereas Schachtel is predominantly concerned with perception. Like Piaget, Schachtel makes no use of Freud's concept of the primary process. He seems to be concerned only with a certain category of phenomena, namely those which are conscious or easily accessible to consciousness. He is right in reaffirming that psychological life aims not only at need reduction but also at sustaining an interest for objects; however, an enduring allocentric interest in an external object must be accompanied by the existence of an inner object—that is, by a mental representative of the external object, which consists of high forms of cognition.

My own work in the field of cognition began with papers on future anticipation[2] and on schizophrenic thinking,[3] which appeared in 1947 and 1948. When I started my studies on cognition I was very much influenced by the psychodynamic interpersonal approach of Harry Stack Sullivan. Later I felt that although the Sullivanian approach was of fundamental importance and should remain a major part of one's psychoanalytic practice and theory, it lacked adequate consideration of the intrapsychic; i.e., the interpersonal dialogical approach had to be integrated with the predialogical or the intrapsychic. However, the study of the intrapsychic, especially in relation to cognition, should not be concerned exclusively with energetics, but with forms and their development.

Form means a psychological structure which may belong to various levels of organization, from the simplest to the most complicated. Forms include perception, imagery, memory, language, thinking, conceptualization, etc.

By development we mean temporal unfolding. My study of development has been to a large extent influenced by Heinz Werner and his school.[37,38,39] Three types of development can be distinguished. The first is the phylogenetic, or unfolding of a psychological mechanism or form through the evolution of the species. The second is the ontogenetic or the unfolding of a mechanism or form through the maturation of the individual. The third type of development is the microgenetic. As this type is less known than the previous two, some words of explanation are required.

Microgeny, as illustrated by Werner,[37] is the immediate unfolding of a phenomenon, that is, the sequence of the necessary steps inherent in the occurrence of a psychological process. For instance, to the question, "Who is the author of Hamlet?" a person answers "Shakespeare." He is aware only of the question (stimulus) and of his answer (conscious response), but not of the numerous steps which in a remarkably short time led him to give the correct answer. Why did he not reply "Sophocles" or "George Bernard Shaw"? How did he reach the correct answer? There are numerous proofs that the answer was not necessarily an established and purely physical or neuronic association between Hamlet and Shakespeare, but that an actual unconscious search went on. In fact, if the same question is asked to a mental patient (either affected by cerebral arteriosclerosis or by schizophrenia in a stage of advanced regression) or to a person who is very sleep or drunk or paying little attention, he may reply "Sophocles" or "George Bernard Shaw." These are wrong but not haphazard answers, inasmuch as they refer to playwrights. The mental search required by the answer had at least reached the category of playwrights. The numerous steps which a mental process goes through constitute its microgenetic development (Arieti[5]).

These three developments, phylogeny, ontogeny and microgeny unfold in time, although with great variation in the quantity of time: from periods as long as geological eras in phylogeny to periods as short as fractions of a second in microgeny. What is of fundamental importance is that the three processes tend to use the same structural developmental plans. We do not mean literally that microgeny recapitulates ontogeny and that ontogeny recapitulates phylogeny, but that there are certain formal similarities in the three fields of development and that we are able to individualize schemes of highest forms of generality which involve all levels of the psyche in its three types of development.

The two aspects of the psyche, the organization of forms (a logical order) and the threefold development (a temporal order) are equally important. Inasmuch as one tends to permanence, the other to change, a double functionality results which constitutes a main characteristic of the psyche.

A process is a psychological mechanism studied both in its forms and at least one of its developments. When the development is ontogenetic or microgenetic, the process may be subdivided into two types, which are what Freud called the primary and the secondary processes. Any process which has reached the secondary stage, has gone through the primary.

When a process goes through a sequence of stages and forms, as they usually evolve in the average human being, we have a normal psychological process. When sequences of stages and forms combine to produce harmful or undesirable effects we have the psychopathological process. When sequences of stages and forms combine to produce something new and useful we have the creative process.

Whole and Part

A topic which is of great pertinence to the whole field of cognition, from the simplest perception to the most complicated conceptual thinking, is the problem of part versus whole.

Under the influence of the Gestalt school, psychology has been dominated by the idea that objects are apprehended as wholes. This point of view is now gradually recognized as incorrect. It was the result of the fact that perception was studied as it occurs in the secondary process and in a state of full consciousness. The study of early ontogeny and microgeny reveals that objects are first perceived as parts and that only subsequently are they perceived as wholes. However, in microgeny part-perception is very rapid and remains unconscious, so that we are aware only of whole-perception, or of the gestalt, which becomes the dominant one at least in the early stages of the secondary process. Riesen,[25] in mammals raised in darkness, and Von Senden[34] in adult humans who were operated for congenital cataracts and were learning to see, have demonstrated how part-perception preceded the total perception of the object. A serial apprehension of the parts preceded perception of the whole. The shortest time in which a congenitally blind person approximated normal perception of wholes was about a month.

In the psychoanalytical literature, Melanie Klein[18] has given great importance to the apprehension of part-object in infants. According to her at first objects which are perceived (or introjected) are partial objects. For instance, the mother is seen as several disconnected parts, often only as a breast. Pritchard, Hebb and Heron,[20] by stabilizing the retina, have reproduced a pregestaltic perception in which wholes are fragmented or part-perception is the only one possible. In tachistoscopic and other subliminal experiments often only parts are registered (Werner[37]). Some parts, which in subliminal experiments are apparently not perceived, were instead registered, as demonstrated by the fact that they later appeared in dreams (Fischer et al.[9,10]). Neurological reports may be interpreted as suggesting that part-perception precedes whole-perception (Alexandra Adler[1]).

Ethology and animal psychology give many examples of animal responses which can be interpreted as responses to parts and not to wholes. Tinbergen[32] has illustrated some very interesting mechanisms; for instance, certain fish, like the sticklebacks, in fighting competitor male fish which invade their territory, react only to the red color of their belly. The color red elicits in the sticklebacks a given fighting behavior towards males of the same species. Since they react blindly only to a part of the total environmental situation and neglect the other parts, they make what seem to us mistakes. For instance, artificial red objects placed by the experimenter elicit the same fighting response. All of us are familiar with similar examples. Moths respond only to the element light by being attracted and may burn themselves to death in contact with powerful electric bulbs. The flypaper, which was used until not too long age to kill flies, exploited part-perception. The fly which reacted only to the sweetness of the paper was caught by the stickiness of it and met death.

It would seem thus that early in phylogeny the organism reacts predominantly or compellingly to *a releasing element* or to a part, and only in a very weak fashion to the rest which is experienced as indifferentiated or as a nonorganized aggregation of parts. It could also be that at very primitive levels there are no such things as parts and wholes, but only releasing elements.

Actually this problem cannot be separated from the problem of the response to the part versus the response to the whole. It was implicit in what we have said that two ways of responding exist in nature: to the part and to the whole (or to the releasing element and to the whole). Each of these two types of responses is generalized; that is, the organism tends to respond in the same way to different exemplers of the same stimulus, irrespectively of whether the stimulus is what we consider a part or a whole.

Generalization is a process applicable to a plurality of stimuli: a plurality of what seems to us parts or a plurality of what seems to us wholes. Generalization of responses to a plurality of objects, which have only a part in common, will eventually lead, at the level of thinking, to the formation of primary classes.[5,6] Generalization of a response to wholes leads eventually to the formation of secondary or logical classes.

Elsewhere I have described how the first type of generalization corresponds to Pavlov's generalization of responses. The second type of generalization corresponds to the Pavlovian technique for discrimination. I have also shown how the responding to parts and not to wholes gives rise to those phenomena described by Klüver about equivalent stimuli, that is about stimuli which appear to human beings very different from each other and yet evoke the same response in some animals. In effect, part-perception and response to a part belong to the primary process whereas whole perception and response to a whole belong to the secondary process.

The Preverbal Levels of Cognition

One of the simplest mental constructs which eventually leads to the formation of the concept is the exocept, or what neurologists call the engram. The exocept leads not to thinking but to external behavior. It is the type of thinking which we use when we play tennis, do manual work, go swimming, etc. It corresponds to what Piaget calls the sensori-motor intelligence. Although, as Piaget has demonstrated, this type of cognition is the prevailing one only in the first two years of life, it has been the almost exclusive interest of behavioristic psychology.

If exocept formation or motor behavior is inhibited, we have expansion and development in another direction: the development of imagery. The image is an internal quasi-reproduction of a perception. Although we cannot deny that images exist in some subhuman species at a rudimentary level, they are predominantly a human characteristic. If by "symbol" we mean something which stands for something else which is not present, the image then must be considered one of the first symbols. For instance, I close my eyes and I visualize my mother. My mother may not be present, but her image is with me; it stands for her. My mother acquires a reality which is not dependent on her physical presence.

Images constitute the foundation of inner or psychic reality, which in human psychology is as important as, and in some aspects more important than, external reality. Cognition does not only help the individual to understand the world, but also to create a surrogate for it. It is by virtue of inner dealings with this internalized surrogate that man can understand the world and see his place in it. The image is the first substitute for the external object. It helps us to maintain an affective disposition for the object, even when the latter is absent. The image is actually an inner object, although not well organized.

The endocept is an intermediary stage between the image and the concept, something similar to what the workers of the Wurzburg group called imageless thought and the Freudians unconscious or preconscious thinking. It is a primitive organization of memory traces, images and exocepts, which take place when imagery and motor responses are inhibited.

The endocept cannot be transformed into a verbal expression. It remains at a preverbal level. Although it has an emotional component it does not expand into a clearly felt emotion. It remains an atmospheric feeling, an undefined intention, an holistic experience which cannot be divided into parts or put into words. At other times there is no sharp demarcation between endoceptual, subliminal experiences and some vague primitive experiences. In many endoceptual experiences there is some awareness, but it is dim and has a global or diffuse quality, which does not lead to any fine discrimination.

Patients in psychoanalytic treatment often mention endoceptual experiences which occurred in early childhood, from the age of two to four, before the acquisition of adult language. Similar experiences also occur in dreams in which verbal expressions do not enter or play a secondary role. The dreamer experiences the dream in a very rich, intense way and yet without words entering into it. When the dreamer is asked by the analyst to describe what he felt in the dream, he uses many words of which he was not conscious while he was dreaming. For instance, he could say that while he was dreaming he was experiencing a mixture of joy, fear and anxiety, as he has never experienced in life. The experience was endoceptual and could only be reproduced very imperfectly by words.

I cannot describe other characteristics of the endocepts, nor of the paleosymbols, which are the first symbols, other than images, occurring at this preverbal level. It is a phantasmic world, consisting of representational phantasies, like images, or by imageless or endoceptual experiences. It is the foundation of that inner reality, which will endlessly enrich and complicate the human situation.

Let us imagine that the subsequent levels that we shall describe in the rest of this paper have not yet developed in early hominids and that a primordial individual has developed only to this preverbal level. This individual would have great difficulty in distinguishing images, dreams and paleosymbols from external reality. He has no language and cannot tell himself or others "This is an image, a dream, a phantasy, an imagination. It does not correspond to external reality." He will tend to confuse the psychic reality with the external reality. Whatever is experienced becomes true, by virtue of the fact of being experienced. Not only consensual validation from other people is impossible, but even intrapsychic or reflective validation is not available. This level of development is characterized by what Baldwin, in relation to very young children, called adualism—that is, absence of the ability to distinguish between the two realities, that of the mind and that of the external world. In dreams and in many cases of schizophrenia there is a return to this state of adualism. This phantasmic world corresponds to a large extent to that inner world to which the British school of psychoanalysis, under the leadership of Melanie Klein and of Fairbairn, have given great importance in their interpretation of man's childhood and of psychopathological conditions.

There is another important aspect of the phantasmic world which has to be taken into consideration. It is a world without causality. The individual cannot ask himself why do certain things occur. He either accepts them as just happening in a sort of naive acausalism or he expects things to take place, in a certain succession, as a sort of habit, not as a result of causality or of an order of nature. The only phenomenon indirectly connected with causation is a subjective feeling of expectancy.

Images with pleasant connections urge the individual to search for the equivalent external objects which are gratifying. Thus, especially unpleasant images and paleosymbols tend to remain as inner objects. The result is that inner life at this level tends to be predominantly unpleasant, a fact already stressed by Melanie Klein.

Preconceptual Levels of Thinking

A higher level of cognition is characterized by the acquisition of language and the paleological or prelogical way of thinking.

One of the most primitive forms of thinking is the one which I have called the primary aggregation,[5] which corresponds to what Schilder[33] called the "sphere" and Linn "the cluster."[19] The primary aggregation does not have a definite holistic organization, except one of general and loose togetherness. It differs from the endocept inasmuch as it contains verbal elements and can therefore be externalized. This externalization consists of the verbal expressions of an element of which it is composed. This element comes to represent the whole primary aggregation. At times this primary aggregation, which is the result of individual experience, seems a strange agglomeration of disparate things, put together in a sort of "collage." At other times it may be considered like an embryonic structure from which conceptual structures eventually emerge. In some cases it may be recognized as including a field which at higher levels of development corresponds to a highly abstract concept. Each element of the primary aggregation may become equivalent to any other element or to the whole aggregation.

It is impossible here to quote more than one example of primary aggregation. The most typical examples can be obtained from primitive languages, some aphasic patients and some regressed schizophrenics. A regressed schizophrenic of whom I asked "Who was the first President of the United States?" replied, "White House." Actually George Washington never lived in the White House, but *White House* was an element of a primary aggregation of cognitive elements which had to do with Presidents of the United States.

Thinking does not generally remain arrested at the level of primary aggregation, but the phylogenetic, ontogenetic or microgenetic developments continue to follow their course. Thinking becomes more and more organized to reach eventually that order of organization which is generally called Aristotelian logic.[3,4,6] I have called *paleological* (which etymologically means ancient or archaic logic) the stage which precedes the Aristotelian. Although paleological cognition is illogical according to Aristotelian standards, it is not haphazard, but is susceptible of being interpreted as following an organization, or logic of its own. This organization consists of connections made between different elements, which eventually lead to the formation of classes. Paleological thinking is well understood in the formulation of von Domarus' principle,[33,34] which, in a formulation slightly different from the original, states: whereas in mature or secondary process thinking identity is accepted only upon the basis of identical subjects, in paleologic (or primary process) thinking identity is accepted upon the basis of identical predicates.

The most typical applications of von Domarus' principle occur in schizophrenia. An example that I often quote is that of a patient who thought she was the Virgin Mary. Asked why, she replied, "I am a virgin, I am the Virgin Mary." The common predicate "being virgin" led to the identification of the two subjects—the Virgin Mary and the patient. The common predicate becomes the *identifying link.* Following is another example. A red-haired young woman in a post-partum schizophrenic psychosis developed an infection in one of her fingers. The terminal phalanx was swollen and red. She told the therapist a few times, "This finger is me." Pointing to the terminal phalanx, she said "This is my red and rotten head." She did not mean that her finger was a representation of herself, but, in a way incomprehensible to us, really herself or an actual duplicate of herself. This type of thinking is also found in normal young children, especially from one to three and a half years of age.

The paleological level of thinking can be interpreted with formulations which apparently differ from von Domarus' principle, but actually refer to the same phenomena. We can interpret the paleological level as one at which special classes are formed. We shall call these classes primary classes because they belong to what Freud called the primary process. We shall call secondary classes those which belong to an Aristotelian level of thinking and to Freud's secondary process. A secondary class is a collection of objects to which a concept applies. For instance, Washington, Jefferson, Lincoln, Roosevelt, Truman, etc., form a class of objects to which the concept "President of the United States" applies. A primary class is a collection of objects which have a predicate or part in common (for instance, being virgin), and which, by virtue of having such predicate or part in common, become identified or equivalent. Whereas the members of a secondary class are recognized as being similar (and it is actually on their similarity that their classification is based), the members of a primary class are freely interchanged. Not only do they all become equivalent, but one of them may become equivalent to the whole class.

It is easy to understand why it is so. In primary classification it is only the common element which counts; all the rest is not important or not noticed or reacted to by the organism. The common element is the only part (or in some cases, predicate) to which the organism readily responds. Perhaps it is this propensity for a response to a part which occasions the assembly of different subjects to form a primary class. In some cases if the response is inhibited the cognitive process proceeds to the formation of a secondary class.

Now, at this point, it is possible to see the relation between part-perception, as described earlier in this paper, and the formation of primary classes. It is also possible to see the relation between whole-perception and the formation of secondary classes. It seems that the human mind operates under the two above-described forms of generalization, both at the level of perception and at the level of conceptualization. What in primary generalization is the releasing element corresponds to what in paleological thinking is the identifying link.

The principle of von Domarus and the formation of primary classes are typical also of the cognition of dreams. Symbolic thinking, as Freud describes it in *The Interpretation of Dreams,* may be seen as following the paleological level of cognition. The Freudian symbol is not only something which stands for something else, but also something which has at least a common characteristic (predicate) with the thing it symbolizes. A person may dream that a king is talking to him. Analysis of the dream will reveal that this king stands for the father of the dreamer. Here it is the similarity of a concept (person in authority) which leads to the identification of father with king. After this paleological identification has taken place at a conceptual level the thought is perceptualized and a visual image of a king appears in the dream. A similar mechanism occurs in hallucinations.

Whereas the phantasmic world was either imageless or predominantly visual, the paleological world is predominantly auditory. It is a world of transformation. Things undergoing metamorphoses are easily identified with other things with which they have just one or more characteristics in common. This is the mythical world which the philosophers Giambattista Vico and Ernst Cassirer have vividly described.

As we have already mentioned in the first part of this chapter, these transformations are interpreted by the Freudian school as mobility of cathexes. Cathectic energies would be easily shifted from appropriate objects to inappropriate ones. It is obvious now that it is the primary aggregation or the paleological type of cognition which permit transformation of objects and the transferring of the affect to equivalents. Even if this presumed cathectic energy shifts to various forms, it is because these various forms exist. At the exoceptic or motor-sensory intelligence level such transformation or shifting would not be possible. At the paleologic level of cognition causality is teleological and not deterministic.

Conceptual Levels

Conceptual thinking as a force in psychodynamic psychiatry and psychoanalysis has been unduly underestimated. The emphasis has been on the primitive, on bodily needs, instinctual behavior and primitive emotions, which can exist without a cognitive counterpart or with a very limited one. Undoubtedly physiological or proto-emotional states such as hunger, thirst, fatigue, cold, sexual urges, fear about one's physical survival are powerful dynamic forces and lead to conflicts and psychological unrest, but by no means do they include all the emotional factors affecting man. Now, in the middle of the twentieth century, much fewer men than ever before starve for food or sex, and yet psychological malaise and mental illness have increased, not decreased.

The most important emotional factors, which motivate or disturb men, are those which are sustained or actually engendered by complicated cognitive symbolic processes. To think that these cognitive processes are only displacements, detours or rationalizations which cover more primitive instinctual drives, is a reductionist point of view. Thus, even Rapaport,[23] who reaffirmed the importance of thought in psychoanalysis and emphasized that drive dynamics translates itself into behavior through thought processes, had an incomplete vision of the problem. Rapaport did not see the other side of the coin, namely that cognitive processes themselves create emotional situations and thus become important dynamic forces.

Schachtel is correct in stating that man is motivated not only by desire to reduce his needs, but also by interest for the object. However, by object we should not necessarily mean external or finite objects, but also inner objects. Inner objects are at times emotional-cognitive representations of external objects; at other times they are representations of concepts and of clusters of concepts. Concepts and organizations of clusters of concepts become depositories of emotions or creators of new emotions and therefore become in their turn great emotional forces. They have a great deal to do with the conflicts of man, his frustrations, his state of happiness or despair, of anxiety or of security. The concept-feelings of personal significance, of self-identity, of one's role in life, of self-esteem could not exist without complicated cognitive constructs.

Faith, loyalty, patriotism, vanity, search for despotic power, pride, hate, aesthetic rapture, love in the agapic or fullest sense, and anything usually referred to as pertaining to the spirit, would not be possible without concepts.

Concepts of concepts and symbols of symbols are continuously created, so that there is no end to the potential conceptual life of man. If this cognitive life may lead to an inexhaustible enrichment, it may lead also to disorders. Concepts may be in conflict and may lead to the psychological dissonance, that Festinger[8] has illustrated: that is, to lack of inner mental consistency.

It may lead to much more serious conditions. In my opinion severe anxiety reactions, schizophrenia, paranoid states, etc., have very little to do with hunger, thirst or sex *per se,* but very much to do with the conflicts arising from the complicated conceptual world of man. In some of these conditions the disintegration of the conceptual world brings about a reversal to a preconceptual level of cognition and may re-establish the supremacy of the phantasmic and paleological worlds of inner reality.

REFERENCES

1. Adler, Alexandra: Disintegration and restoration of optic recognition in visual agnosia. Arch. Neurol. & Psychiat. 51:243-259, 1944.
2. Arieti, S.: The processes of expectation and anticipation. Their genetic development, neural basis and role in psychopathology. J. Nerv. & Ment. Dis. 100:471, 1947.
3. ——: Special logic of schizophrenic and other types of autistic thought. Psychiatry 11:325, 1948.
4. ——:Interpretation of Schizophrenia. New York, Brunner, 1955.
5. ——: The microgeny of thought and perception. Arch. Gen. Psychiat. 6:454, 1962.
6. ——: Studies of thought processes in contemporary psychiatry. Am. J. Psychiat. 120:58, 1963.
7. Arlow, J. A., and Brenner, C.: Psychoanalytic Concepts and the Structural Theory. New York, International University Press, 1964.
8. Festinger, L.: A Theory of Cognitive Dissonance. Stanford University Press, 1957.
9. Fisher, C.: Dream and perception: The role of preconscious and primary modes of perception in dream formation. J. Am. Psychoanalyt. A. 2:380-445, 1954.

10. ——: Subliminal and supraliminal influences on dreams. Am. J. Psychiat. 116: 1009-1017, 1960.
11. Freud, S.: Collected Papers, vols. 1, 2, 3, 4, 5. London, Hogarth, 1948.
12. ——: The Interpretation of Dreams. New York, Basic Books, 1960.
13. Gill, M. M.: Topography and systems in psychoanalytic theory. *In* Psychological Issues vol. 3, New York, International University Press, 1963.
14. Hartman, H: Essays on Ego Psychology. New York, International University Press, 1964.
15. ——, and Kris, E.: The genetic approach in psychoanalysis. Psychoanal. Study Child 1:11-29, 1945.
16. ——, and Lowenstein, R. M.: Comments on the formation of psychic structure. Psychoanal. Study Child 2:11-38, 1948.
17. Jones, E.: The Life and Work of Sigmund Freud, vol. 1, New York, Basic Books, 1953.
18. Klein, Melanie: Contributions to Psycho-Analysis. London, Hogarth Press, Ltd., 1948.
19. Linn, L.: The discriminating function of the ego. Psychoanal. Quart. 23:38-47, 1954.
20. Pritchard, R. M., Heron, W., and Hebb, D. O.: Visual perception approached by the method of stabilized images. Canad. J. Psychol. 14:67-77, 1960.
21. Rapaport, D.: Organization and Pathology of Thought. New York, Columbia University Press, 1951.
22. ——:Toward a Theory of Thinking. *In* Organization and Pathology of Thought, New York, Columbia University Press, 1951.
23. ——: *In* Medical Psychology, International University Press, 1953. P. Schilder, ed. New York,
24. ——: The Structure of Psychoanalytic Theory. New York, International University Press, 1960.
25. Riesen, A. H.: The development of visual perception in man and chimpanzee. Science, 106:107-108, 1947.
26. Schachtel, E. G.: On memory and childhood amenesia. Psychiatry, 10:1-26, 1947.
27. ——: Metamorphosis. New York, Basic Books, 1959.
28. Schilder, P.: Medical Psychology. New York, International University Press, 1953.
29. Spiegel, Rose: Specific problems of communication in psychiatric conditions. *In* S. Arieti, ed. American Handbook of Psychiatry, vol. 1, 909-949, New York, Basic Books, 1959. pp.
30. ——: Communication in the psychoanalysis of depression. *In* J. Masserman, ed. Psychoanalysis and Human Values, New York and London, Grune & Stratton, 1960.
31. Tauber, E. S., and Green, M. R.: Prelogical Experience. An Inquiry into Dreams and other Creative Processes. New York, Basic Books, 1959.
32. Tinbergen, N.: The Study of Instinct. Oxford, Clarendon, 1951.
33. Von Domarus, E.: The Specific laws of logic in Schizophrenia. *In* J. S. Kasanin, ed. Language and Thought in Schizophrenia: Collected Papers, University of California Press, 1944.

34. Von Senden, M.: Space and Light: The Perception of Space and Shape in Congenitally Blind Patients Before and After Operation. London, Methuen & Co. Ltd., 1960.

35. Weinstein, E. A., and Kahn: Denial of Illness: Symbolic and Physiological Aspects. Springfield, Charles C Thomas, 1955.

36. ——, and ——: Symbolic Reorganization in Brain Injuries. *In* American Handbook of Psychiatry, vol. 1, S. Arieti, ed. New York, Basic Books, 1959, pp. 964-981.

37. Werner, H.: Microgenesis and Aphasia. J. Adnorm. & Social Psychol. 52:347-353, 1956.

38. ——: Comparative Psychology of Mental Development. New York, International University Press, 1957.

39. ——, and Kaplan, B.: The development approach to cognition: Its relevance to the psychological interpretation of anthropological and ethnolinguistic data. *Am.* Anthropologist 58:866-880, 1956.

THE DYNAMICS OF INFORMATIONAL ADJUSTMENT PROCESSES

By JAMES G. MILLER, M.D., Ph.D.*

WHILE REVIEWING the progressive development of Freud's theories of mental process in his distinguished biography, Ernest Jones indicated that cybernetics might in the future contribute to such formulations.[10] Jones understood, as intimately as anyone, Freud's views and his continual, restless efforts to revise and improve them. Possibly, if he were alive today, Freud might have recast his thought in the light of the recent developments in cybernetics, information theory, and systems theory. We can never know the validity of such a speculation, but it may be worthwhile from the vantagepoint of the mid 1960's to review some behavioral researches with a systems orientation, in order to see whether they apply in any way to modern psychoanalytic thought.

Fundamental Systems Concepts

A system is a set of units with relationships among them. This definition covers living systems, which belong to the class of "concrete systems," as well as abstracted and conceptual systems. Concrete systems are organized, nonrandom accumulations of matter-energy in regions in physical space-time, made up of coacting, interrelated subsystems. Living systems are a special subset of concrete systems which includes plants and animals, all of which are organizations of protoplasm, sharing certain fundamental characteristics of life among which is keeping specific critical functions within appropriate ranges.

General systems behavior theory conceptualizes these systems as occurring in a hierarchy of different "levels," each of which has as its components the systems of the level beneath and each of which, in turn, helps to compose the levels above it. All systems have subsystems

*The author is grateful for support for this work from a grant from the Carnegie Foundation and from Grant MH 08607 from the National Institute of Mental Health.

and are components of suprasystems. Cells, tissues and organs, organisms, groups, social organizations, societies, and supranational systems are the chief levels of an ascending series of living systems.

All living systems are open systems. They combat entropy by receiving inputs from outside their boundaries, carrying out internal processes making use of these inputs, and transmitting outputs across their boundaries. All living systems process both matter-energy and information. The sets of subsystems which carry out these two sorts of processing are in many ways comparable. Matter-energy processing, which includes the intake and use of food, water, air, and other such inputs, is outside the scope of this paper. Just as there is a metabolism of matter-energy, so there is also a metabolism of information. Man is the living system we are concerned with here, but comparable processes exist in other species of organisms, and in systems at all other levels.

Bits of information reach man's input transducers (specialized sense organs). Within the system, information is transduced by "internal transducers" or proprioceptors. The information then passes, through shorter or longer nervous and hormonal channels and nets to the decoder, which changes the messages from the public language outside the system boundaries to a private language of the system. The information is then combined with other learned information and stored in memory. Some of this information is used in the output which crosses the system boundaries into the environment or suprasystem. A central decoder brings together information from all matter-energy and information-processing subsystems, determining what will be transmitted on for information output. This is then processed through an encoder, which translates it into the public language. An output transducer turns nervous impulses into speech and gestures.

Information which is transmitted at rates or with meanings which the organism can easily handle requires no special adjustment processes. Inputs, however, can be either so great in amount or so pregnant with meaning as to constitute excess stresses (either of matter-energy or of information), or so slight as to lack stresses. For such variations special adjustment processes are available. A stress may arrive at the system boundaries unheralded or it may be preceded by a warning "threat"—prior information which makes possible adjustments aimed at avoiding the stress altogether or reducing its seriousness. The "steady state" of systems is the resultant of adjustments by all of a system's subsystems to a never-ending stream of inputs, and to controlled changes in state within and among the subsystems.

The many matter-energy adjustment processes of human organisms, because of space limitations, cannot be discussed here.[12] They include much of the subject-matter of various special fields such as biochemistry, physiology, and medicine. Adjustments to information flows can be categorized as input, internal, and output adjustments.

Information Input Overload and Underload

We have experimentally studied adjustments to information input overload—a rate of input beyond the easily handled load—at the levels of cell, organ, organism, group, and social organization. At the level of the cell, we stimulated single neurons of frog sciatic nerves by electric pulses separated by different brief temporal intervals. The ouput was recorded farther down the same cell. Making the assumption, on the basis of various sorts of evidence, that information in this neuron is coded by frequency modulation rather than amplitude coding, we could calculate the maximum number of bits of information per second that could be output by such a neuron as we put in information at various rates. We calculated that the neuron could keep up fairly well with the input up to about 4000 bits per second. Above this rate the output first leveled off and then began to fall, ending far below the peak rate of 4000 bits per second.

We next tested the visual pathway of the rat. The optic nerve was stimulated photically and electrically, and the maximum possible output calculated from data collected in the area of the calcarine fissure of the occipital lobe. This organ showed a comparable performance pattern to that of the single cell except that the peak capacity was only 55 bits per second.

At the levels of organism, group, and social organization we used human subjects. The tasks used required pressing levers as rapidly as possible in response to briefly presented light flashes. The flashes were presented in pairs with different interstimulus intervals of 50, 100, 150, 200, 250, 300, and 350 milliseconds. At the organism level, each subject saw both flashes in each pair and was instructed to respond to each by pushing a lever as soon as possible after he saw each flash. At the group level, one subject saw the first flash of each pair, one saw the second, and each pressed a lever as soon as he saw his flash. These lever pushes, in turn, produced light flashes seen by the third member of the group, who responded to each of the two flashes by pushing his lever as soon as he could and releasing it. At the organization level, we used a similar procedure with seven subjects arrayed in three echelons. Two physically separated two-man groups were the input transducers. Both members of the first group pushed levers in response to the first light flash in each pair. Both members of the second group responded similarly to the second light flash. Their responses in turn flashed stimuli before an intermediate two-man echelon of information processors. One man saw two flashes coming from the first group in the first echelon, and the other saw the two flashes coming from the second group in the first echelon. When he saw two flashes, each man pressed a lever. The responses of the two men in the intermediate echelon produced light flashes seen by the one man composing the third, output echelon. He responded to each of the stimuli presented to him by pushing a lever.

Results in these studies produced curves similar to those found for the cell and organ. Human subjects operating as individual organisms followed the input rate of information fairly well up to about 4 pulses per second (250 milliseconds interval), transmitting a maximum of about 5.75 bits per second, but at faster speeds they began to lag behind. The groups had similar curves, following the input up to about 3 pulses per second (330 milliseconds interval), transmitting at most about 3.5 bits per second before beginning a gradual decrease. The organization reached a channel capacity of about 2.5 bits per second (400 milliseconds interval), and then decreased slowly toward zero.

In the course of meeting the input overload a number of adjustment processes are used by living systems, some by all and others only by the more complex systems. A human individual may use omission—leaving out part of the information; error—transmitting false information without correcting it; queuing—storing part of the information in memory until there is time to return to it; filtering—setting up a priority system so that parts of the message considered more important are transmitted first and the rest later; generalization—reducing the information to an abstract or summary; and escape—refusing to process any further information.

Each of these methods of adjustment has its costs, measured in time, energy, loss of reward, or other units. Ordinarily a person will first use the adjustment which appears least costly to him. It is pathologic to do otherwise. Information input overload is a common stress in our complex society, particularly among executives and others who must make decisions based on large amounts of information. Techniques for handling this flow can be learned and should be available to executives.

Information input underload also can be a severe stress, particularly at the level of complex organisms such as human individuals. Such stress can also arise out of life situations. Explorers, prisoners, truck drivers traveling alone, and people isolated in respirators or alone in rooms as a result of illness may be affected by it. It has also been studied in the laboratory with subjects in experimental situations in which all sensory input is restricted for relatively long periods. Responses to this sort of situation vary with the degree of deprivation, the length of time the person must endure it, his personality, and his motivation. In both life situations and experiments, increased fantasy, even to the point of hallucination, panic, and psychotic-like states have been found to occur.

Adjustments to information input underload also exist. They can maintain a reasonably steady state over varying periods before they fail and hallucinations begin. Increase in fantasy, for instance, provides information from within the organism from stored memories and associations. Monkeys deprived of light press a bar for a light reward that enables them to see the informational patterns of their environments. This response increases as deprivation becomes greater.[6]

Human subjects, given the chance to relieve the underload conditions by pressing buttons for flashes of light, pressed more frequently for series of lights containing more information than for those with relatively little.[9] Lilly reported wiggling his fingers and toes when under input underload stress.[11] A prisoner, in solitary confinement in a Chinese communist prison, made up for input underload by spending hours daily searching through the information in his memory.[14]

Anxiety

"Anxiety," like pain and discomfort, can be interpreted as an information signal. Anxiety is probably a signal of the relative strength of a threat or stress and of the adjustment processes which are available to counter it.

Freud believed that anxiety was related to pleasure and pain. He held that the modification of heart action and respiration at birth arouse the first anxiety. He distinguished between objective anxiety, a condition of increased sensory attention and motor tension which acts as a signal of anticipated danger from without, and neurotic anxiety which appears purposeless. Anxiety is a signal and it elicits adjustments to the dangerous situation it signals by either flight or defense. If it does not, past experience takes control and the resulting affective state is paralyzing and unadaptive. Freud concluded, that what one fears in neurotic anxiety is one's own libido. "The difference between this and objective anxiety lies in two points—that the danger is an internal instead of an external one, and that it is not consciously recognized."[8]

Pavlov also explained anxiety, which he observed in his experimental animals.[13] He considered anxiety to arise from a conflict between excitatory and inhibitory processes. The animals showed it by motor excitation, whining, and abnormal breathing. In ordinary circumstances, an equilibrium is set up in which the animal is quiet and reacts to stimuli by excitatory or inhibitory responses. Sometimes, however, the conflict disturbs normal activity so that it goes on for days or years. The animal may return to normal after the experiments are discontinued or may require treatment.

Freud and Pavlov have expressed their views, but the present evidence seems to suggest that anxiety is a signal, like pain, of a lack of steady state somewhere in the system. Under threat or stress, the signal is transmitted when a comparison of the relative strengths of the stress and the available adjustment processes indicates that the former may be stronger. Either an increase in the threat or stress or a decrease in the adjustment processes as perceived by the organism can produce this signal. Like pain, anxiety is a useful signal to organisms. It can also be pathologic, becoming attached to inappropriate objects and circumstances, or the threshold for it can become lowered to the point where it is provoked by too slight changes in the steady state.

Experiments on Anxiety and Stress

A number of experimental studies make use of physiologic measures to determine if it is true that threat and stress evoke anxiety. Urinary steroids, which increase after stresses eliciting pituitary-adrenal response can be measured accurately. Changes in electrical resistance of the skin, resulting from increased output from sweat glands which accompanies anxiety, and muscular tension, said to increase with anxiety, can be measured. Psychologic tests, such as the Taylor Manifest Anxiety Scale correlate positively with subjectively reported feelings of anxiety.[17] Frequently, however, in studies of anxiety, threat and stress, definitions of terms are not clear and it is sometimes difficult to be sure what is being measured.

Berkun, Bialek, Kern, and Yagi subjected military trainees to a number of situations likely to arouse anxiety.[1] In one of these they were told that the plane in which they were riding was about to "ditch"; in others that they were caught by a forest fire or trapped within the range of artillery fire; in another that their actions had caused serious injury to a man. In some of these situations tasks were assigned—e.g., sending messages on a radio which required "repair" but which was actually irreparable. Urine and blood samples were taken to determine the changes brought about by such situations. Changes in eosinophil level of the blood and steroid content of the urine were found in response to many of these situations. The subjects' behavior showed a number of adjustments. A few subjects gave way to their fear and ran. Most obeyed instructions and worked on the radio, although some of their behavior was maladaptive. For example, when they believed they were threatened with artillery fire, although the outside temperature was 100 to 125 degrees, they put on fatigue shirts and field jackets as a sort of armor. Once they began work on the radio, danger-appraisal took second place to task-oriented behavior. They rationalized the position at the radio as one of greater safety. Each shell burst supported the feeling that the shells were not going to hit that position although they were detonated in an approaching pattern. Interviews following the experiments showed that the men made use of various ego defenses, some of which made it possible to remain in the situation and go on working, whereas others did not.

Fox, Gifford, Murawski, Rizzo and Kudarauskas[5] have carried on a continuing study of human stress, using psychiatric patients during manic activity, during emotional disturbance, and in the course of psychoanalysis. They found evidence of pituitary-adrenal arousal under these circumstances. They also studied the Harvard rowing team under the stress of competitive rowing and found that racing obliterated the usual morning-to-evening decrease in 17-hydroxy-corticoid excretion and caused an increase during the race when compared to the same period on a day with no competitive rowing. Practice rowing did not increase it. The degree of integration of the team, better one year than another, importantly affected the amount of stress experienced by individuals since the more integrated team suffered less stress. Clearly, informational stress can elicit comparable hormonal reactions to those elicited by tissue damage, although experimental results show that it is less reliable and in smaller degree. When stress is long-continued, according to one point of view, psychosomatic illness can result. Another point of view regards symptoms as adjustment processes in themselves. Anna Freud's list of defenses of the ego includes symptom formation.[7]

Those who consider psychosomatic diseases to be by-products of stress reactions usually hold that "conditioning factors" determine which bodily component will break down under the stress. Theories emphasizing the informational causation of psychosomatic illnesses often assume that there are personality differences, among the people with such diseases, relating to particular sorts of psychosexual developmental experiences. Cohen attempted to put to an experimental test the commonly stated view that asthmatics and hypertensives differ in their levels of aspiration.[4] A group of asthmatics, a group of hypertensives, and a control group were studied in their responses to a task in which, after each trial, they had to state their "level of aspiration" for the next trial. Behavior in response to success or failure was also studied. Although initially it appeared that there was a striking and statistically reliable difference between asthmatics and hypertensives in the direction in which they differed from the controls, they could not replicate the experiments on other patients, so until now this question remains unanswered.

Output Adjustments

In addition to the input and internal adjustment processes considered above, there are output adjustments. Adjustments in rates and amounts of secretions or excretions, in amplitude and intensity of motor behavior, as well as in the part of the motor apparatus and the number of components of it used for output, are among the changes in matter-energy output which can result from informational as well as matter-energy processing. Like the matter-energy output adjustments, alterations in output of information can follow changes in the internal adjustment of the organism. Information can be suppressed or distorted so that output messages are inappropriate to the situation. Information may be coded in a different language to adjust to the speech of another suprasystem. Rate, intensity, and form of output may all be changed as a result of the operation of adjustment processes.

Feedbacks

The cybernetic process of feedback—in which part of the output from a system is channeled back, becomes input to it, and aids in control of further output—is useful to explain certain aspects of the behavior of living systems. Within the organism, components and subsystems are controlled by feedback loops which adjust them to varying needs of the whole system. For instance, posture and skilled motor performances are guided by such signals from motor outputs fed back through input and internal transducers. Stark has shown in precise detail how autonomic feedback controls the size of the pupil in the eye and consequently the amount of light reaching the retina.[16]

Feedbacks also control more complex behaviors. Tracking tasks are good examples. These involve a closed feedback loop including an operator who adjusts dials or moves a lever to compensate for changes in the position of a spot of light, a needle, or some other signal. Tracking behavior is often studied to make quantitative determinations about an organism's input-output relationships.

Chase, Cullen, Openshaw, and Sullivan, for instance, investigated the effects of differences of the form of visual feedback on patterns of control of index-finger movements.[2] A beam of light was displayed to each subject, which moved in a manner directly proportional to the movement of the finger in space and made any error in the movement immediately apparent. The "gain" of the error signal was varied from 1 to 2, 4, 10, 20, or 40 times the actual movement. These six display conditions were presented in random order. Different movement patterns resulted from different conditions. Increasing the gain produced a sharp drop in error.

Delay of information feedback results in disturbed performance. Chase, Harvey, Standfast, Rapin, Sutton, Gilden, and Guilfyle studied the effects of delayed auditory feedback on speech and on motor behavior, involving the tapping of a key.[3] In the speech experiments, delay was provided by having the subject wear earphones and speak into a microphone. His speech was recorded on tape and then, after a brief lag, played into the earphones. The tapping of the key caused a click, and this also could be electronically delayed. They found that delayed auditory feedback caused subjects to increase the intensity of each response, increase its duration, decrease the rate of responses, and increase the number of erroneously repeated responses. When a number of responses was required, delayed feedback increased errors in this number. These experimenters also examined the effects of delayed visual, auditory, tactile, and proprioceptive feedback on key tapping. On a task which required that a regular rate of tapping be maintained, delayed feedback did not have much effect. When patterned ouput was required, such as tapping in groups of three taps each, only the delayed tactile feedback failed to disturb the performance. Increased intensity and decreased rate of tapping resulted from all the others, as well as from delayed combined sensory feedback. Under all delayed conditions, four instead of three taps tended to be produced. The most marked disturbance came from alterations in proprioceptive feedback. These experiments show how essential unimpaired feedback is for proper behavioral adjustment.

Another important type of feedback control is revealed by the effect of performance results on "level of aspiration." In such experiments, a subject is asked what results he expects to achieve on a problem, such as a test of skill, of memory, or of problem-solving. It is commonly found that success in reaching this criterion causes the subject's expectations to rise on further tests, whereas failure causes his expectations to fall.[15]

In all such feedback studies, the fine control of system activities by inputs derived from the system's own outputs is shown.

Subtle social interactions, like maintenance of "appropriate distance" from others in space, "appropriate tone" of voice, or "appropriate aggressiveness" in conversation also may well be maintained by such feedbacks.

Conclusion

This brief review suggests that Jones may have peered prophetically into the future when he associated cybernetics and psychoanalysis. It appears that not only stresses, such as information overload and underload, but also many other informational adjustment processes, including those which depend on feedbacks, can be related to many of the aspects of personality which modern psychoanalysts are concerned with in their psychotherapeutic work.

REFERENCES

1. Berkun, M. M., Bialek, H. M., Kern, R. P., and Yagi, K.: Experimental studies of psychological stress in man. Psychol. Monog.: Gen. & Appl. 76: (Whole No. 534), 1962.
2. Chase, R. A., Cullen, J. K. Jr., Openshaw, J. W. and Sullivan, S. A.: Neurocommunications unit, NIMH. St. Elizabeth's Hospital, Washington, D.C.
3. ——, Harvey, S., Standfast, S., Rapin, I., and Sutton, S.: Studies on sensory feedback. I. Effect of delayed auditory feedback on speech and key-tapping. Quart. J. Exper. Psychol. 13:142-152, 1961.

   ——, Rapin, I., Gilden, L., Sutton, S., and Guilfyle, G.: Studies on sensory feedback. II. Sensory feedback influences on keytapping motor tasks. Quart. J. Exper. Psychol. 13:154-167, 1961.
4. Cohen, L. D.: A note on the repetition of experiments. *In* Psychiatric Research Reports, American Psychiatric Association. Members of the Committee on Research, eds., J. S. Gottlieb, Chairman, 1955-56, pp. 39-42.
5. Fox, H. M., Gifford, S., Murawski, B. J., Rizzo, N. D., and Kudarauskas, E. N.: Some methods of observing humans under stress. *Ibid.*
6. Fox, S. S.: Self-maintained sensory input and sensory deprivation in monkeys: a behavioral and neuropharmacological study. J. Comp. Psychol. 55:438-444, 1962.
7. Freud, A.: The Ego and the Mechanisms of Defense. New York, International University Press, Inc., 1946.
8. Freud, S.: New Introductory Lectures on Psychoanalysis. New York, Norton, 1933, pp. 113-117.
9. Jones, A., Wilkenson, H. J., and Braden, I.: Information deprivation as a motivational variable. J. Exper. Psychol. 62:126-137, 1961.
10. Jones, E.: The Life and Work of Sigmund Freud, vol. 1. New York, Basic Books, 1953, pp. 369-370, 393.
11. Lilly, J. C.: Learning motivated by subcortical stimulation: the start and stop patterns of behavior. *In* Reticular Formation of the Brain, H. H. Jasper, ed. Boston, Little, Brown, 1958.
12. Miller, J. G.: Living systems. New York, Wiley & Sons. In preparation.
13. Pavlov, I. P.: Relation between excitation and inhibition, delimitation between excitation and inhibition, experimental neurosis in dogs. Experimental Psychology and Other Essays. New York, Philosophical Library, 1957, p. 233.
14. Personal communication.
15. Sherif, M., and Sherif, C. W.: An Outline of Social Psychology. (Rev. ed.) New York, Harper, 1956, p. 51.
16. Stark, L.: Stability, oscillation, and noise in the human pupil servomechanism. Proceedings of the IRE, 47:125-139, 1959.
17. Taylor, J. A.: A personality scale of manifest anxiety. J. Abnorm. Soc. Psychol. 48:285-290, 1953.

HOLISTIC THEORIES OF NEUROSIS

By EUGENIA HANFMANN, Ph.D.

In 1965, Wiley and Sons will publish *Neurosis and Treatment, a Holistic Theory* by Andras Angyal, who in his earlier book (*Foundations for a Science of Personality,* The Commonwealth Fund, 1941) presented a conceptual framework for personality study. Angyal was acutely aware of the large number of neurotic patients who had not been helped by years of intensive treatment; he was tireless in his search for more effective methods and intent on determining, from empirical evidence, the essentials of the therapeutic process and the conditions of its success and failure.

Angyal's *basic theories* avoided abstractions as absolutes and did not conceive processes as static structures; instead, with W. Stern he advocated "psychophysically neutral" concepts (such as system, set, shift of set) applicable to all levels of personality functioning. Mental states are symbolic elaborations of the life process, but they reflect only parts of the total process, and sometimes reflect it in a distorted manner. Yet they are not mere reflections: they influence and potentiate the organismic processes they reflect. Angyal uses a single term, "biosphere," for the totality of the subject-object interactions to indicate that person and environment are not two separate entities, but two poles of a single process; life.

Angyal discerned a trend toward increased *autonomy* and one toward *homonomy;* in his book on neurosis he refers to them also as mastery and love. In the one attitude the human being strives to impose his control on the environment; in the other, he seeks to participate in something larger than himself—a relationship, a group, a cause, an ordered universe. The two trends, though seeming opposites, are but two directions of one organizing process: while the individual fits environmental items into his own life as its parts, he also extends his life by becoming a part of larger meaningful wholes. In most behavior both trends are present but one usually predominates; a person can strive, e.g., to influence nature to serve his needs, or he can "resonate" to it in an aesthetic experience. Values and norms of behavior prescribed by the culture are adopted by the individual not only for fear of punishment or ostracism; they can function also as genuine expressions of his homonomous trend, and in this case their violation gives rise to real guilt, not to neurotic guilt based on fear.

Angyal pointed out that the double pattern of life operates also on the physiological level, where processes serving the continuation of the species can be distinguished from those serving the individual, but he saw no reason to ascribe greater reality to the physiological than to any other aspect of the organism's functioning. In choosing to view the specific in terms of the general, as an aspect or a part of a whole, he rejects the reductionist aspects of psychoanalysis, though he respected its clinical viewpoints and findings and incorporated them into his work. Granting that the physiologically fixed functions are dynamically very important, Angyal still views them as only some of the many manifestations of personality, not a basis from which everything derives; the concept of id has no place in his system. The basic tenet of the holistic approach is that the parts must be viewed in the light of the organizational principles governing the whole. To provide conceptual tools for dealing with organized wholes, Angyal outlined the logical properties of systems, as distinct from relationships.

Angyal defines systems as distributions of members in a dimensional domain governed by a single principle. Members or parts of a system function as such only through those of their qualities that are relevant to the system principle; they are connected with each other only through their common participation in the system. For this reason, causal thinking, which looks for direct connections between antecedent and subsequent events, has only a limited role in personality study. System dynamics, as distinct from atomistic causality, are governed by the trend toward a more complete realization of the system principle; the Gestalt laws of "tendency to closure" and "tendency to Praegnanz" are expressions of this trend. Some of the subsystems of personality are formed by a rearrangement of the same part functions to form different total functions: the same muscle contraction may at different times form a part of widely different motor patterns. Multiple functioning—e.g., multiple motivation of a single act—is the rule in personality dynamics, but this economic feature can become a source of disturbance. When one part belongs to two or more subsystems, competition for this part between these subsystems may lead to their mutual interference and endanger the integrated personal functioning.

We turn now to Angyal's *theory of neurosis.* In his conception neurosis is not a focal disturbance, a "bad" part in an otherwise healthy organism: neurosis is a system, and therefore a sweeping condition and a complete way of life. This is why an established neurosis is not easily overcome. It is an organization, sustained and perpetuated by the principles of system action, and it cannot be dislodged by any segmental partial changes. How does this organization come about and how is it different from the healthy organization of personality?

A well functioning person invests his energy in a direct vigorous pursuit of the basic human trends. By his mastery over the environment and creating and maintaining affectional ties he participates in the world in which he lives. By contrast, neurosis is a partial *isolation* from the world, and—since life is interaction—a state of impoverished narrowed-in life. Anxiety is the reflection in consciousness of this state of affairs, the person's evaluative response to a threat to human existence. Should isolation become complete and permanent, the exercise of both mastery and love would be made impossible. This is why the experience of anxiety contains an anticipation of total destruction, in the absence of any realistic danger to life.

Isolation is experienced by every child when, after birth, it gradually begins to develop an awareness of its separateness from the world or from its mother, a differentiation between the "I" and the "non-I." The early image of the outside world is constructed by the child from experiences which disrupt his comfortable unified state and from those parts of the field which—unlike his hands or his legs—do not obey his wishes; the world so constructed is marked by its *alienness* and its tremendous size and power. It is a threatening world, but normally the child overcomes the anxiety resulting from his separateness by exercising the basic human trends; he finds his place within the world by learning to master the environment and, most importantly, by establishing *communication and community* with others through loving relationships. If the child is given ample opportunities from the start to explore and influence his surroundings and to experience closeness with others, by the time the independent existence of the outside world is firmly established he will be safely anchored in it, not isolated and helpless. In such favorable conditions isolation and anxiety are present only minimally, as unavoidable correlates of individual existence, and of growth and change.

If during the crucial developmental period the child's opportunities for achieving mastery and for experiencing love are reduced through some environmental factors, his state of isolation persists; the world that can be neither mastered nor communicated with remains alien and threatening. Soon a turn in the development occurs which leads to the formation of a nuclear neurosis. Instead of continuing his direct attempts at relating himself to his environment, the child now diverts a large part of his efforts toward protecting himself from the world. As he secludes himself within a protective wall, the real world, with its varying opportunities for love and mastery, becomes more forbidding than before; he has to develop ever stronger defenses and more circuitous ways of approach. The process goes on until neurosis reaches the point of relative stability, providing the person with a limited domain in which he can function after a fashion. This development is accompanied and promoted by certain implicit fantasies about oneself and others which come to shape everything the person thinks, feels and does. The self is felt to be small and weak, the world is overwhelming and unmanageable; it also appears cold, and one finds oneself unloved, but even this is not the worst. In time the child begins to feel not only unloved but unworthy of love; this secret conclusion, which undermines even the hope that anyone could ever love him, marks a crucial step in the development of neurosis.

But since it is equally impossible to renounce the basic human strivings, the neurotic must attempt to live and exercise his basic functions in a world in which—according to his own assumptions—it is impossible to be effective. This paradoxical enterprise is possible only because the expectation of doom is not held by the neurotic with absolute certainty: phenomenologically, there is always an element of doubt in anxiety, and correspondingly also of hope. Neurotic life with all its complexities is built on this flicker of hope—hope that it may be possible to love, possible to govern, even though it must be done in tortuous roundabout ways. In contrast, healthy orientation is based on an attitude of sound confidence: the person feels that he is able to help himself and has enough to offer to others, so that he is likely to be accepted. In relation to the environment he feels that though there is plenty of trouble in the world things usually can be worked out one way or another. The basic neurotic assumptions, which are usually masked by pseudonormal beliefs, contrast strongly with these moderately optimistic attitudes. The neurotic's anticipation of dangers and his inadequate twisted maneuvers result in rebuffs and failures;

his dire expectations are confirmed. He dreads the crucial developmental changes because to meet them adequately he would have to break out of his protective confines. The goal of therapy is to transform this limited stationary life into a forward-moving one.

Central to Angyal's theory of neurosis and therapy is the thesis postulating a *dual organization of personality,* or, to use his term, the theory of *universal ambiguity.* Personality develops simultaneously around two nuclei and forms two patterns, one of which may be underdeveloped but never absent. Confidence and diffidence, conviction and doubt that life is livable, mark the "great divide," the point at which the human life course acquires its basic existential conflict and its dual organization. Health and neurosis are two dynamic gestalts organizing the same material, each according to its own system principle, and competing for dominance; the person is neurotic or healthy depending on which system is dominant. Each personality process has a position within these two different patterns and, similar to a part of a visually ambiguous gestalt, changes its character when a shift occurs and figure and background change places. Every personal event has a double meaning and function, depending on whether it takes place within the gestalt of health or the gestalt of neurosis; hence universal ambiguity.

The conception of personality as a dualistic organization, with one of its two patterns predominating, has far-reaching consequences: many major concepts must be reappraised and reformulated in the light of this theory. Defense mechanisms, e.g., are viewed by Angyal as organizational devices, specific types of system action, through which each system seeks to complete and maintain itself consistently with its system principle, thereby preventing the alter system from reaching a position of dominance. The systems organized and defended by these mechanisms are not the conscious and the unconscious, or the ego, superego, and id; the struggle for dominance takes place between the two incompatible orientations, two major ways of adapting to life. "Defense mechanisms" can serve either system, they are a two-way affair. Thus both the neurotic feelings and trends and the healthy ones can be repressed, the latter by the neurotic organization. Therapy aims at reinstating to dominance the latent system of health. Resistance is the self-defense of the neurotic pattern which becomes stronger as the threat to neurosis increases. Angyal finds no place in his conceptual framework for the concepts of id and ego, viewing the latter simply as a collective name for the organizational devices in personality. However, superego is accepted by him as a distinct subsystem of personality which can form a part of either of the two major organizations and change correspondingly. He has a great deal to say about the duality of the superego, its healthy and neurotic aspects and roots, and about how and why neurotic guilt and real guilt have to be handled differently in therapy.

In describing the early traumatic origins of the neurotic pattern, Angyal is quite explicit about the role played by cognition. He defines *trauma* as a persistent interference with the person's basic healthy pattern of life, which cannot be managed by him in accord with this pattern and eventually leads to its distortion. But whether or not a given event becomes traumatic depends also on the child's "view" of the event, which in turn is codetermined by his developmental level and by his level of competence and confidence. Objectively harmless situations can function as threats and can produce major traumas in the absence of ways to cope with their disturbing effects. Furthermore no event can be considered traumatic unless its effects on later behavior are generalized beyond the particular situation in which they originated. *Generalization* is a precondition of the formation of the neurotic pattern.

According to Angyal, generalization is present whenever a single object or event is seen as representative of a much broader condition or issue. Being disappointed by a person in whom one had absolute trust may destroy not only the specific belief that this particular person is trustworthy but the general belief in the goodness of people. An adult need not be overwhelmed by such an experience since he has a capacity to distinguish between the specific and the general, but this is not the case with the young child whose experience is global and undifferentiated. For the young child, the home and the people in the home are more than the representatives of the world: they actually are his *whole* world; the specific *is* the general. Since the mother fills the world, her actions teach the child that such is life and such is the world. If he feels rejected, his feelings, translated into adult language, are not "my mother rejects me" but "God has forsaken me, humanity has cast me out." If he is hemmed in by taboos he feels that the world is against him, that the exercise of his will is punishable as such.

If traumatic experiences occur before the child's world has expanded and become differentiated, "generalization" does not occur *after* the event, separately from the cognition of the concrete situation; it is built into the experience and creates a pessimistic pattern of expectations not just with regard to the mother but to life as a whole—not a partial but a total pattern. Negative experiences made in later stages, being partial patterns, need not have effects of the same magnitude. But the early formed negative pattern will continue to determine the child's total view of the world even after his life has expanded beyond its early confines. Suppose a child who had felt rejected later develops a well founded conviction that his mother actually loves him; but now she is just one person among many, and her love cannot undo the rejection that came from her as the primal mother, equated with the universe. This incongruity is one of the main reasons for the failure of continued attempts to solve neurotic problems. For example, in seeking the affection of a particular person the patient is driven by the hidden wish for reconciliation with the primal parent, for acceptance by the one who was all in all. But if the concrete goal is attained, the affection received necessarily proves to be that of one particular and imperfect person, which leaves the patient with the feeling, "Is that all?" He is likely to devalue the relationship and to renew his search.

Angyal's conception of the role of generalization in the genesis of neurosis determines several features of his therapeutic approach. One of them is his emphasis on *holistic interpretation.* Since neurosis is not a collection of specific disturbances but a system of generalized attitudes, significant improvement will result only from changes in attitudes of wide scope. Consequently, interpretive work should focus not on single symptoms or traits but on the persisting broad attitudes which are indicated by the totality of the reported and observed manifestations. By focusing on specifics and their origins the therapist may play into the hand of the neurosis, since the patient himself often resorts to the method of focalization in order to ward off the more vital general issue.

Angyal quotes the case of a patient who had such a strong aversion to birthmarks that he would promptly reject any person on whose skin he could discover the tiniest freckle. His former therapist let him spend much time speculating about this symptom, e.g., the possible connection between birthmarks and the color of the nipple, but—as the patient told Angyal with poorly veiled triumph—they had failed to solve the riddle. This man in his daily life showed an unusual degree of neurotic fault-finding; he would quickly find defects in most people and situations and declare them completely worthless. While he was greatly concerned with the hidden significance of birthmarks, he was not at all concerned about the obvious destructiveness of his life.

According to Angyal, the specific symptom should be used, in conjunction with many other similar manifestations, for bringing to the patient's awareness his most pervasive neurotic attitudes with all their destructive results. The first step consists in gathering into larger units items that can be viewed as instances of the same theme, though the same item can represent more than one theme. Holistic interpretations lead from specific items to syndromic units and through them to personal attitudes of wider scope. This personal "neurotic mythology" has to be outlined before one explores its origins in the early circumstances of the patient's life. The tracing of origins involves also the demonstration that the neurosis has some roots in the person's pattern of health. To make therapy successful we must complement the uncovering of the neurotic structure by a careful unearthing and rehabilitation of the repressed healthy patterns.

Angyal's conception of the nature and consequences of early generalization is essential also for the understanding of the role he ascribes to *transference.* According to him, the great intensity transference can reach is not due to an emotional mistaking of the therapist for one of the family members as they were known after the child's world has come to include several people; this most powerful transference arises from investing the therapist with the significance of the "primal parent" and consequently experiencing him almost as one's whole world. Transference in this sense is manifested less in the specific reactions to the therapist than in the overwhelming emotional importance which for a time he acquires for the patient. If the patient does not experience the therapist as representing the "world" whose acceptance he craves, or succeeds in quickly divesting him of this function, he will explain away or discard as unimportant the acceptance he has received, or any other potentially corrective experience; the specific as such cannot affect the general. But interpersonal experiences made in the state of strong transference may have a powerful effect because their all-inclusive totalistic character matches that of the early experience. If an interpersonal situation is created by the patient which repeats an early traumatic one, and this situation is given a different outcome by the therapist's response, the effect may be almost miraculous; the patient's attitude to the world at large may be changed. By the same token, if the therapist is led by his own involvement or lack of acuity to respond inadequately to a crucial situation, the effect may be tremendously magnified by transference; the patient may literally re-experience the earlier trauma in its full force, and his neurosis may be perpetuated.

Even if such dramatic occurrences are infrequent, this description makes obvious the importance of the therapist's behavior; the positive or negative quality of his attitudes is more often than not correctly perceived by the patient and merely magnified by transference. Although Angyal fully acknowledges the value of the analysis of transference for the uncovering of the neurotic structure, he emphasizes even more strongly the resconstructive value of the real relationship between the patient and the therapist the nature of which depends largely on the therapist's attitudes as they manifest themselves in the sessions. The therapist's role is not merely that of a stimulus for transference: he functions also as the representative of health for the patient and can be a powerful influence in mobilizing him for change, by acting impersonally he forfeits this influence.

To what extent is therapy a cognitive endeavor, and cognitive in what sense? Angyal leaves no doubt about the importance of cognitive restructuring which takes place in the first part of therapy, but he is no less emphatic in stating that insights as such, i.e., as mere statements of fact and of their connections, cannot bring real changes. If the patient is to discard the neurotic pattern and, in the face of intense anxiety, essay a new and unfamiliar way of life, he must have *experienced* his neurosis with its destructive effects, and also experienced in some small measure his suppressed individual pattern of health. A full experience of an issue includes, besides purely cognitive moments, an evaluation of its personal relevance which is embodied in emotions. These are concrete experiences which convey vitally values of which thoughts are merely a pale reflection. Insight can pave the way to change only if, seeing the state of affairs differently, the patient also starts *feeling* differently about it.

It follows that the therapist's function can not be limited to fostering insight: *facilitating experiences* is an even more vital task. The most important messages are not conveyed through explicit formulations, and one does not interpret experiences and feelings that one wants the patient to keep. Angyal emphasizes that the therapist must learn to communicate in implicit ways, dramatically, when necessary. The patient's progress may depend on the therapist's ability to maintain the sessions at a high level of emotional tension. Nevertheless, Angyal warns us not to ascribe therapeutic value to emotional responses as such, since emotions, as all other personality processes, may express either neurosis or health. As long as the neurotic pattern is strongly dominant the patient values all that maintains and furthers it, abhors and fights off all that is contrary to it. The expression of these neurotically founded emotions is productive in therapy only insofar as it clears the way for other feelings, or results in further insights. Even potentially useful functions such as catharsis are easily perverted into the service of neurosis. Real progress is in sight when the patient starts responding to his intellectual discoveries in an increasingly adequate manner, no longer rooted in a hopelessly negative image of himself and the world, or in the fantasies meant to ameliorate or disguise it.

Such vital realizations serve to bring the patient to the point where he is ready to face the issue of change, and when he has reached this point insights have served their purpose. After recognizing his problem in its full extent the patient must confront it as the central issue of his life: he is in a crucial predicament and must do something about it. The second half of therapy consists not in continued insight gaining but in the struggle to discard one's old attitudes and let the new ones take over. The central neurotic issue, the choice between confidence and doubt, trust and distrust, stagnation and movement, is decided in a global manner, but the struggle for decision may be intense and protracted. It may culminate in despair when the patient realizes the bankruptcy of his neurotic ways of life, yet is held back from a radical change by anxiety. However, if the old patterns are sufficiently weakened, the patient's field of vision is extended to a realm of new possibilities of feeling and acting, and a shift toward health takes place. As the patient redefines himself and his world, the neurotic alternatives recede into the background and become ineffective. They are not however completely abolished; since neurosis is a pattern of organization, not a localized growth, it cannot be "extirpated." Relapses are therefore a common occurrence; Angyal advocates studying the conditions under which they occur as well as the factors and methods enabling the patients to regain their health and remain well.

COGNITIVE THEORY AND PSYCHOPATHOLOGY*

By HARLEY C. SHANDS, M.D.

INTEREST IN HUMAN THINKING is not new; indeed, if we consider that thinking in symbolic terms is the specifically human attribute, we find from the very beginning of recorded history that philosophers have been preoccupied with this problem. Perhaps even earlier, we can find poetic evidence in the story of Narcissus that in man's discovery of *reflection* he discovers himself while falling in love with an apparently external image. In this context, we may note the interesting contrast of approach in two systems: Freud characteristically adapted the mythological story of Narcissus in a metaphorical explanation of certain phenomena of man's reflexive relationships, whereas Piaget in his own abstract manner uses the idea of reflection as the basis for the final stage in the maturation of human intelligence.

By general agreement the beginnings of modern science and modern philosophy can be traced to the work of Descartes, and it is perhaps specifically of interest that Descartes's most famous preoccupation is in the cognitive area. From the Cartesian doubting the pervasive doubting of scientific method directly emerges; much that is current has a very close relation to Descartes's decision to place all his reliance upon observation of thinking. From Cartesian origins we also see in current application the ever-present danger threatening any theoretical system dealing with these complex matters. The danger is that of *personification,* that all-too-human tendency to revert to the animism with which the primitive human being peoples the sky, the land, and the sea with mythological deities, nymphs, and centaurs.

In the immediately current scene we see an extension of the Cartesian

*This investigation was supported in part by U.S.P.H.S. Grant No. MH 05518-03 from the National Institute of Mental Health and in part by a grant from the Commonwealth Fund.

The quotations from Emily Dickinson used in this article are from *The Poems of Emily Dickinson,* edited by Martha Dickinson Bianchi and Alfred Leete Hampson, and appear through the kind cooperation of the publisher, Little, Brown & Co., Boston, 1939, to whom thanks are extended.

doubting method to even that single undoubted entity which he found to establish a center for his own life. So distinguished a critic as Bertrand Russell points out that the observation of thinking does not necessarily imply the existence of a personified thinker. The observation of thinking certainly indicates that thinking is going on, but this is an observation which in modern language can be considered a behavioral observation; what we begin with is the observation of certain predictable regularities, and the recourse to personification in explanation is clear evidence of a lack of comprehension of the mechanics by means of which the behavioral observations may be abstractly construed. In personification in any context, we find a close resemblance in technique to the mythological explanatory system: in myth, to explain wind, one postulates a wind-god blowing. What personification demonstrates in the modern sense is a strong and regular tendency repeatedly evident in man's early attempts to grasp conceptual techniques. As Freud himself mentions in one comment upon his own personifications, the notion of "instincts" in the libido theory is a mythological notion. It perhaps needs to be emphasized that modern understanding of the symbolic process indicates that the process itself has a definite "life cycle," and that mythological explanations are the essential early steps in moving toward an abstract understanding; Cassirer in particular emphasizes that mythology is the childhood of language.

The contrast in approach is well seen in comparing the poetic beginnings of a theoretical system, which appear so intriguingly in Freud's writings, with his expert grasp of the way in which classic human myths demonstrate in poetically transparent form an early stage of human psychological process, and especially the manner in which myths like that of Oedipus and that of Narcissus demonstrate the unconscious human understanding of the human unconscious long before it becomes possible to describe the unconscious in conscious terms! In one way, we repeatedly find that advances in human knowledge regularly follow this process, namely that of finding means of expressing in conscious, discursive terms the insights formerly graspable only in the metaphors of myth and fantasy.

It is then of the greatest interest to note the counterpoint of the systems of Freud and Piaget, since in Piaget's approach we find a formidably abstract system stripped to its intellectual bones, and, perhaps because of this parsimoniousness of statement, difficult of comprehension. In contrast to the familiarity and warmth with which we can greet many of the Freudian metaphors, if we are to follow Piaget, it is clear that we must be prepared to take a long journey without much help along the way. But the end of such a journey is well worth the trip, since from the Piaget system we derive understanding without the crutches of personification; crutches which, like other kinds of crutches, support but enfeeble.

The two systems of Freud and Piaget stand in many ways in sharp contrast to each other, in approaches which are mutually complementary but often quite contradictory. What joins them in spirit perhaps more than any other feature is that both are derived from direct observation of behavior in the situation of interest to the investigator. Freud's interest in patients is similar in approach to Piaget's interest in children; where Freud extends his interest in his patients to his own dreams and neurotic manifestations, Piaget extends his interest in children to the detailed examination of his own children in infancy. The immense difference in approach in this clinical method is sharply in contrast to the classical method of philosophy, that of introspection unsupported by detailed observation of behavior in a systematic sense. Where both Freud and Piaget have illuminated our understanding by close attention to the similarities between the intimate and the public observables, philosophers in the older tradition tended to develop *Weltanschauungen* as the products of a self-fertilizing introspective romance. The benefits of exogamy are well shown in the systems of Freud and Piaget in contrast to the results of intellectual incest, even in the case of Descartes!

Another approach to be considered (see chapter by J. G. Miller) is that of communication engineering, and it is noteworthy to see some points of similarity and difference in this manner of looking at the cognitive process. Where both Freud and Piaget are system-builders, and in each case the system is clearly identified by and associated closely with the person of the founder, the situation in communication engineering is quite different. Here we find that although there have been and still remain many notable figures in the field, the whole is a composite product rather than a single systematic effort. In this way, the approach of communication engineering shows a much closer relation to other disciplines in modern science. In another way, communication engineering is pre-occupied throughout with technical problems. Where in Freud's approach the relation between theory and technical operation is often remote, so that in psychoanalysis it is often good practice to discuss no part of the theory with the patient, and where even in Piaget's system the data must be translated into theoretical terms by a process of extended interpretation, in communication engineering the relation of theory and practice is very close indeed.

Communication engineering is more than clinical, it is directly and immediately practical. In many ways, the approach of communication engineering is typically American, with strong emphasis upon gadgetry and instrumentation in contrast to the more reasoned and academic approach of the European traditions from which both Piaget and Freud take origin. In modern applications, communication engineers are faced with a greatly accelerated process of mutation and selection having some resemblance to the process of biological evolution. Testing in application accelerates obsolescence. In sharp contrast to this immediacy of test with selection or rejection in a relatively brief period, we see that the systems of Piaget and Freud each went through a very prolonged period of growth and development before proving to be of value to a wide audience.

The difference is perhaps most evident in the importance of mechanisms. Where Freud postulated "mechanisms of defence" and where Piaget sees "intellectual operations," both in a rather abstract manner, in communication engineering techniques the mechanism has to be susceptible to design and concrete production of the material for testing. Theories have had to be tailored to the direct applicability of mechanism to instrumentation. It is therefore all the more exciting to find in this field many instances in which the practical evolution of methods to approach problems of communication gives rise to mechanisms later found to be replicas of mechanisms developed in biological evolutionary process.

Let me just cite two examples of these trends. In the first place, in a brilliant insight, Boole developed a system of logic in the nineteenth century on the basis of a binary system of notation. With extraordinary foresight, he called the book he wrote, which remained long obscure, *The Laws of Thought.* From this beginning, modern logicians developed the modern field of symbolic logic, the field so important in Piaget's recent work. The remarkable development of the computer mostly in the past two decades is traceable in part to the discovery that data could be processed much more efficiently if expresed in terms of a binary system of notation. But, in the process, returning to an examination of the nervous system in its mechanical operations, it was then found that neural data-processing operates in similar fashion, on the basis of a frequency code of binary form. (Granit, 1955) The more recent application of ideas from communication engineering is the dramatic discovery of the methods of genetic transfer of instructions from one generation to the subsequent one. Here again, transmission is by code, and the analogy to the "instructions" or "program" fed into a computer is very close indeed. It is perhaps of greatest interest, in terms of early mythological beginnings, that we find here too a process in which with great effort we discover, outside, shapes which in an indirect way reflect biological human processes.

In an effort to present in a very condensed summary some extraordinary recent implications in human knowledge of human function, let me first construct, then systematically destroy, a metaphor which illuminates the process both of scientific investigation and of the human mentational function. The formation and destruction of hypotheses follows the sequence in which human beings in their earliest fumblings reach out for mythological explanations to support understanding until it becomes possible to substitute for the myth a more abstract hypothesis. In human history, in fact, we find that every epoch is occupied with the destruction of the myths of its predecessors, and with the construction of new ones to be eventually destroyed in their turn. The second fact that can repeatedly be observed is that the construction of any consistent system of myths is so powerful a factor in reducing human uncertainty that there is an inevitable tendency to establish belief around these integrated myths-in-system. The problem then appears that whereas it is relatively easy to substitute belief for uncertainty—indeed, as indicated by the comment, "there are no atheists in fox-holes," the greater the uncertainty, the greater the tendency to achieve a sustaining belief in the system at hand—it is always more difficult to supplant one belief system by another.

This means in practical terms that it is much easier to arrive at a belief than it is to change a belief already established. The reason is easy to find: as Pierce notes, belief is the basis for action, and without belief, or when serious doubt has been cast upon the accuracy of belief, the human being finds himself in a state of doubt and anxiety. The result is that man only reluctantly and with great effort changes a belief system. The process, further, is continuously relative, since even great discoverers, after battling resistance to their own discoveries tend in their turn to resist the next discovery (Kuhn). This process was beautifully described by Max Planck, who then proceeded to demonstrate the accuracy of his own statement by his own later action; in another modern instance, Einstein, after overthrowing much of the intellectual basis for the classical physical theory, resisted the subsequent discovery of the importance of probability in physics by making the statement, couched in properly mythological terms, that he could not believe that God played at dice with the universe!

The little myth which, like a modern "happening" I would like to construct for the sole purpose of demonstrating its destruction, runs like this: Man can be compared with a model-maker who begins the construction of a facsimile of a ship, much reduced, within a bottle with a very narrow neck. With enormous effort and patience, and with the greatest of difficulty, he introduces into the bottle, bit by bit, the components of the "inner world" which in its developed shape reflects and represents the external world. In this model, the neck of the bottle obviously represents the routes through which the components of the eventual model are internalized. The "motivation" assumed to impel the model-maker is that of constructing the inside facsimile as closely as possible similar to the external original, and this is carried out through the process of "representation" in the central nervous system.

The working model so constructed remains not too distant from the idea that many persons in this field still retain. But I assert here and shall support the assertion with references to detailed evidence, that we know in every particular that this model is in error.

The very first assumption can be immediately characterized as completely in error. The notion that the "inner world" is constructed to resemble the outer world in some consciously perceived model must be immediately replaced by another, more complex notion. What is established is rather a probabilistic system of implicit or "unconscious" schemas (cf. Bartlett, 1934 and Head, 1911) which serve as some kind of abstract templates for comparison. The operation of the system is such that any new message received which agrees sufficiently with any stored message (or "expectation") can be ignored—in this mechanism the ground is cleared for the differential apprehension of the unfamiliar (Brazier). What we notice is the novel, and what we deal with competently is the *phenomenally* novel, in contrast to the widespread disturbance in the physiological systems when we encounter the *categorically* novel. The distinction here made between the phenomenally and the categorically novel in "external" terms appears to be reflected in Freud's differentation of respectively signalling "secondary" anxiety, and the disrupting "primary" anxiety. The storage system appears to be similar to that of a computer: what is stored is a coded message, and the information carried in perceptual experience is derived from a comparison—the information is the *difference from expectation.*

The central position of the notion of "representation" in neurology, in psychology, even in psychoanalysis is still current, in spite of evidence which seems to me reliably to contradict even the possibility that this is a correct notion. Representation in this sense appears to be the modern version of a theory embodying the idea of "preformism," and it can be seen in comparison with the long-abandoned idea that the spermatozoon contains a microscopic man who simply grows larger in his development. This latter idea has been abandoned for centuries, but the idea of representation still lingers in other didactic systems. The British neurologist Gooddy recently reviewed the evidence indicating that representation is a formulation which can no longer be entertained on the basis of relevant evidence. We know, for instance, that the brain is functionally the site of circulating nerve impulses, ceaselessly travelling in enormous numbers. Some of these nerve impulses, coming from certain areas of the body, are processed in certain areas of the central nervous system—but the conclusion that the peripheral area is "represented" in some specific part of the brain is no more accurate than to assume that one's house is "represented" in the central telephone office. There are destinations and sources of messages, and there are paths in which the coded impulses travel, but there are no images, no homunculi, no soul, no mind—at least to the best of informed investigation with detailed and delicate modern instruments. There are only travelling patterns of nerve impulses in constant motion, and perhaps some stabilized patterns of facilitation.

Skinner remarks, in a recent paper reviewing fifty years of behaviorism, that the oldest crude errors in human thinking derive from primitive animism. The notions of representation reflect the primitive belief that since human beings experience apparent conscious control, similar processes of control must be features of every entity in the universe. Descartes's postulation of a soul resident in the pineal gland is a somewhat cruder version of the idea of representation in the cerebral cortex. With the help of analogies to modern machines, we find that it becomes possible to accomplish *replication* without having to postulate *representation.* For example, it is possible to project images on a television screen by camera and subsequently to replicate them in response to instructions recorded on magnetic tape. The image is not "represented" in any way on the tape; the situation rather is that the camera behaves by performing an operation of recording, and the tape recorder, by reversing the operations of recording, performs the behavior of displaying. The camera-television tape recorder has the advantages and disadvantages over the eye-brain system of very accurate and literal reproduction, but in the two cases, what happens is the "behavior of seeing" in two different contexts. What is seen is identical, in spite of the difference in origin of the information. The analogy to remembering in human terms appears very clear, but the human machine is so much more complex that processes of deformation and contamination of stored memories begin immediately, in contrast to the much slower, uncomplicated, decay in pattern in the stored tape.

In another important application of the same general notion, it is evident that no "thinker" or "model-maker" is necessary to act as origin for the process. The television camera, the satellite "spy in the sky" and the eye all collect information through the sensitivity of a specially prepared surface and transmit that information by processing it locally to yield a coded message which is then transmitted along some channel to a destination equally prepared to receive the information and to transmit it in turn. The mechanisms, whether biological or mechanical, operate in their own manner because such a method is designed into the mechanism. Whether by the rapid process in which the modern engineer designs a circuit with sensitive elements, or the extraordinarily slow process by means of which similar mechanisms have evolved in the blind processes of organic evolution, it is not necessary to go further than to say that the function depends upon the design. The modern interest in cognitive process is accelerated by the idea that what we are looking for is a comprehensive understanding of the design of the human neural machinery.

To return to the metaphor: it is easier to make the formulation that a model-maker is making a model, inside, to replicate the object, outside. If we, following the reasoning above, abandon the idea of the personified agent carrying out consciously desired behavior (an abandonment for which Freud performed the necessary preliminary spade work) then it follows logically that we may also abandon belief in the metaphysical existence of the "object" outside. In computer slang, we find engineers saying that the machine makes decisions, pursues goals, and controls processes—all functions clearly humanoid in character. But the engineer does not mean that the machine behaves in such a fashion because "it wants to"; it is simply that it is much easier to sum up the capabilities of the whole into a single entity with a name. We see in psychoanalytic theory an application of this same trend in the use of the term "ego," which is generally agreed to be not an entity but simply a designation for an integrated collection of functions. In a similar way, putting the picture the other way round, in a modern rendering, we find that an "object" is similarly to be understood as an integrated idea of the source of regular messages.

The slow development of the idea of object has been brilliantly investigated by Piaget. His painstaking demonstration of the slowness with which "object-concepts" are formed clarifies in a most important way that the notion of a universe of objects, and the objectified self within that universe, is a construction rather than a given.

A close study of cognitive function reveals that the possibility of arriving at the notion of a self appears only at a relatively advanced stage of human development. Such a notion has to be learned, and it can be shown that it is learned in a particular setting. In Western thought, the concept of a self is developed along the same lines as the concept of an object; as Whitehead puts it, the developed conception in the adult in this version is that of self as "Ego-object amid objects," in a spatially homogenous matrix. In his pioneering study of linguistic symbolism, in contrast, Whorf comments that in the Hopi language, no such notion of objectified man, or of homogeneous space, or of kinetic (as opposed to dynamic) time is to be found. The inference is very compelling, to wit, that all of our conceptions of ourselves in the world are in fact learned in the symbolic system of our particular culture.

A paradoxical result of the unconscious process in which the model-maker builds his world is that while building the objects of this world, he also and reciprocally builds himself. The lesson which is perhaps the bitterest for many patients in psychotherapy emerges from this consideration, since all of us have to learn that, although we are, as Housman said, "I, a stranger and afraid, in a world I never made," we have to take the responsibility for this world as though we had consciously and with design constructed it just the way it is.

Therefore, if we cling to any trace of the original metaphorical presentation, we find in the developmental history of the human being the curious event (traceable entirely to the maturation of techniques of what Piaget refers to as reflective thinking) that as the construction of a universe proceeds, the innermost part of this universe is separated out as a distinct institution, the self. A close examination of the process indicates a conclusion stated by Bertrand Russell, to the effect that the primary space of human life is a perceptual space, and that while, from one line of investigation, we are forced to the realization that the brain is a physical object among physical objects in the process of objectification which is basic to science, a full application of what we know indicates without the slightest doubt that the perceptual space is "in the brain" in some essential sense. This process is difficult of statement in scientific terms, because of the linearity of reflective thinking in the discursive mode; but it is easier in the poetic phraseology of Emily Dickinson, who says:

> The brain is wider than the sky,
> For, put them side by side,
> The one the other will include
> With ease, and you beside.

Modern investigation of the nervous system indicates that its basic nature is circular; the "storage" mechanism predicted by Kubie and so effectively demonstrated in later investigation is that of reverberating circuits which preserve the apparent stability of the mentational system by a ceaseless circulation—the process is one of an apparent "perpetual motion" in a recirculation like the fountains in the drugstore window where one sees a faucet endlessly running without apparent source of supply. We face here the remarkable paradox that the basic circularity of process can be grasped in conscious terms only through its expression in the extremely narrow channel of discursive speech and logical analysis.

In this respect we find it possible to return to our all-but destroyed model to take another look, this time to find one relevant implication combined with another seriously in error. The relevant implication is that the channel of communication between inner and outer world is indeed constricted. We know from anatomical and physiological investigation that the nervous system has an incredible capacity, constituted of the possible interrelations of a central mass of more than ten billion central components and the related peripheral extensions; but this vast capacity has to be utilized through the narrow channels of input, and, if we are to deal in *conscious* terms with the world, with the incredibly narrowed input of information couched in verbal or mathematical symbols. The fact that success can be attained in this difficult problem is demonstrated by Helen Keller: human beings, given basic good equipment and the dedicated services of an inspired teacher, can achieve a complete inner world through channels narrowed even further, without the centrally important distance receptors of sight and sound.

When we see the difficulty of the use of the narrow input channels, and especially the pitfalls of verbal behavior, we can understand the paradoxical demonstration that human beings prefer to construct inner entities access to which is rendered immediate by processes of reasoning and recall rather than to rely upon a process of continuous investigation of the apparently external world. The words of the physicist Planck are especially to the point here; after pointing out specifically that "all the ideas employed in physics are derived from the world of sense-perception," Planck goes on to note that considerations (which in the context of Piaget's system are primarily considerations of *conservation*) compel us to assume the existence of another "world of reality behind the world of the senses . . . a world which can only be perceived . . . by means of certain symbols which our sense allow us to apprehend." In this way, Planck expresses a metaphysics of his own, commenting first that we must use sensory data, and second, that these sensory data must be processed in terms of certain symbols which themselves can only be understood as stabilizations of sensory data, to find the "reality" of the highest type. In this method, Planck expresses a faith precisely similar in form to the faith of the religious in the "existence" of a "higher reality." Without in any way contesting the presence or absence of such a higher reality, I would simply like here to underscore that the route to any inner entity is the route of data-input and data-processing, and what emerges must be a function of the process of its construction.

The award of the Nobel prize in physics was occasion for a recent comment by Wigner as to the goals of physics and the limitations of physical conceptions. Wigner notes that physics "does not endeavor to explain nature"; I would emphasize that this statement does indeed say "not." He goes on, "In fact, the great success of physics is due to a restriction of its objectives: it only endeavors to explain the regularities in the behavior of objects." With one qualification, this comment expresses in precise terms what I want to assert here; the qualification which I would make is that the words "of objects" are not only unnecessary but actually misleading. The notion that "objects exist" is a metaphysical notion which reflects the ancient human need to believe in the substantiality of the outside world. While no one seriously questions the significance of this belief in practical terms, it is by now widely accepted that the actual existence of objects is insusceptible of proof. Physicists in modern times have learned to deal with such concepts as "electron" without having to believe in its objective status. The electron of modern physics is the electron which works, not the electron which exists. Therefore, Wigner's statement appears to say enough when it says that physics (or any other science for the matter) investigates *regularities of behavior* which are predictable and reproducible without necessarily achieving a metaphysically "real" status.

The reason I belabor this point with perhaps excessive repetitiveness is perhaps that I too find it very difficult to believe. All human beings want to find a world in which there is the inner certainty of solid objects and replicable events; the point is especially poignant to psychiatrists occupied in their daily duties with evidence that many human beings experience the persistent suffering of not finding a stable world while not being able to adapt their expectational systems to anticipate the inevitability of change. We return to the clinical consideration that in human beings the attitude of anxiety is that attitude in which we confront uncertainty, and the loss of anxiety the predictable result of the construction of any inner world having the appearance of certainty and the support of consensus.

I would like to suggest, with a final reference to the model of the model-maker and his model, that our concern in this symposium is with neither the "external world" nor with the "inner world" but primarily with the channel of relatedness which is in fact the primary reality from which both inner and outer world differentiate in the process of human maturation, with especial emphasis upon the manner in which the symbolic process forces us to deal with objects in space and time. What we find, both in the systems of Freud and of Piaget, is a clear demonstration that the channel is the opposite of passive. It is not a simple bottleneck, it is in fact the site of remarkable changes of sequential nature throughout human life from infancy to maturity to sensescence. It is characteristic of human life that the child's experience should be that of incessant novelty, with, in many places the experience of delight expressed by Wordsworth in the phrase, "apparalled in celestial light," whereas that of the old man should be, in the words of the Preacher, "There is nothing new under the sun." It is perhaps the most central feature of the communication channels of human beings that there should be a progressive failure of capacity to deal with novelty in adult life which follows a progressive increase in tolerance for novelty (or anxiety, its other side) throughout a favorable childhood.

We know from research into the development of neural connections that it is not until the connections are made and the whole matures (with the necessity in some neurons of waiting until myelination is complete before the given function can be adequately trained) that the channel can function as a means for communication. But in the subsequent developmental process, Piaget's studies have demonstrated that there is a timetable for the maturation of symbolic function—it is necessary for a human being with good training to wait until adolescence to become capable of operating in terms of propositional logic and the procedure of hypothesis and test which characterizes scientific inquiry. This means that not only in the physiological but in the pedagogical sense, it is impossible to use a channel for communication which is not adequate to carry the information required. As Miller has reported here, the process, instances of "information overload," analogous in some ways to the overloading of electrical circuits for carrying power, can be demonstrated as both prior to and characteristic of, psychiatric disorder. The developmental processes, emphasized especially in the work of Freud, of Piaget, and in that of our recently mourned colleague Werner, are intimately involved at all times; if we return concretely to the metaphor with which we began, it is evident that the "model-maker" must cope at every successive stage of his development with a channel having radically different characteristics; in a psychoanalytic context, the latency child is abruptly and sometimes explosively propelled into the adult world by the sudden violent maturation of sexual process which leaves him at the very beginning of his adult sexual life more highly endowed than he will ever again be, while necessarily innocent of any "emotional insight" as to how to master and exploit this novel influx of power.

The developmental changes in physiological process and the related changes in cognitive capacity—surely there can be no doubt that the association of maturity of propositional logic and sexual maturity is not mere coincidence—make it inevitable that there should be a great gap in the understanding of the child by the adult and the adult by the child. As suggested by Schachtel in a psychoanalytic idea recently adapted to the understanding of computer technology by Neisser, the problem appears more clear when this basic difference in code is taken into account rather than by the assumption of another "little man in the machine," the censor who lowers the curtain on infantile sexuality in the process of inducing "infantile amnesia." We would find a difference in the ability to communicate which might be similar in kind were we to try to arrange a conversation between one of Babbage's original computing machines of the nineteenth century and the vast monsters of the present day: it seems likely that a modern computer, if able to carry out a process of "agentification," might well postulate that the older machine, fixed in its infantile structure, was "resistant" to modern modes of data- processing approach.

It is in this context, I think, that we must go back and look again at Freud's monumental success and his reasonably understandable failure. Freud was able, at the price of constructing an inner world based upon the metaphysics underlying an already obsolete Newtonian view, to point out (1) that human life is much more than the life of reason, (2) that the "reality" of human relatedness is vastly more effective and subtle, both for disease and for happiness, than is accountable on the basis of the narrowly limited operations of the tiny channel of conscious knowing and (3) that the accumulated treasures of myth and metaphor still have significant meaning in human understanding of the human condition.

The point of central importance is that of relatedness; human life is a matter of the establishment of relationships, between data and nerve impulses at the basic level, between percepts and concepts at a higher level, and between human beings at the highest levels. Because of the limitations in any one human being at any one time in history, it is inevitable that most of what we think we know must be wrong; but it is important that we exchange information among ourselves so that we may come to new syntheses as we climb from one plateau to another with the painful slowness which characterizes the accretion of facts and the inevitable demonstration of the erroneous nature of the system which made it possible in the first place to collect the very facts involved.

Finally, I want to refer again to a poetic insight. Emily Dickinson in this poem describes a paradoxical attitude appropriate to the scientist who must ever look for evidence which negates his most cherished convictions. In Miss Dickinson's words, this investigator must be:

> Magnanimous as bird
> By boy descried,
> To sing unto the stone
> Of which it died.

Really to learn requires a willingness to congratulate the destructive critic who proves us wrong, and to agree joyfully with the "analysis of resistance" basic to a widened understanding of the human condition. To this end, I propose that the destruction of ancient ideas implicit in much modern research into cognitive processes and the mechanisms of thinking is painful only if we have reached a stage in which rigidity is essential to stability.

BIRTH ORDER AND CREATIVE STYLES

By IRVING D. HARRIS, M.D.

THE AIM of this chapter is to submit theory and evidence which suggest that birth order can be fruitfully employed as a rough indicator of basic and divergent psychological tendencies. Although, there are, to be sure, inevitable clinical implications, these are indirect and will not receive major attention. The principal emphasis will be rather on the intellectual repercussions of these divergent tendencies. For this reason, the thesis can be considered as most pertinent to the topic of cognitive or creative styles.

The potential fruitfulness of birth order is indicated by the fact that the order in which one is born into a family has long been an intriguing subject for the unknown makers of myth, the psychologists investigating genius, the psychoanalysts, and even the political scientists. In the Biblical story, birth order and its repercussions underlie the fratricidal conflict between Cain and Abel and the power struggle between Esau and Jacob. Psychologists such as Havelock Ellis and Lewis Terman were interested in what bearing birth order had on the development of creative genius. Sigmund Freud pointed to the importance in a man's biography of his position in regard to brothers and sisters, and Alfred Adler more systematically described the personalities of the oldest, middle, and youngest child. Alexis de Tocqueville, who was astute enough in the early nineteenth century to predict that the United States and Russia would emerge as the two future world powers, mainly ascribed the fall of an aristocracy and the emergence of a democracy to the abolition of the inheritance law which gave the first son the entire estate. Birth order clearly appears then to be no trivial factor in human affairs.

Theoretically, birth order can be looked upon as reflecting the competitive power struggle between siblings—a position taken by Alfred Adler[1] as reflecting variations in parent-child involvement—a viewpoint shared by Freud and more explicitly so by such investigators as MacArthur[12] and Schacter.[15] The classical study of birth order by Helen Koch[10,11] takes into account both viewpoints. The theoretical position I shall develop pertains to the parent-child relationship and, without minimizing the importance of competitive sibling interaction, considers birth order relevant to the broad question of how much was the adult-parent involved in the humanization of the child.

We are accustomed to thinking that the parents are the principal early humanizers of the child. But this is certainly more true in small families than in large ones. An only child would have more opportunity to be humanized or civilized by his parents than would the seventh child in a family of fifteen children. In the latter case, the child's older brothers and sisters would have a prominent part in rearing him. Even in smaller families, say of three or four children, the child's older siblings may be required to take on a semiparental role toward him. Furthermore, there are changes in the parents as more children are born. Whereas with the very first born or the first born male, there is usually an intensity of parental involvement—an intensity compounded of interest, ambition, and anxiety—with those born later the parents are usually more relaxed and accepting or more matter-of-fact.

It can be seen that there may be considerable differences in how intensely the parent is directly involved in the humanization of the child. Although there is, unfortunately, no exact way of measuring these differences, a rough indicator is provided by birth order. As we shall see, the general effect on the child can be described in terms of how soon he adopts the adult viewpoint, the first borns tending to do so sooner, the later borns later. To be sure, the trends described do have exceptions—a younger child coming after a long interval may have his parents quite involved in his humanization. Furthermore, a later born from one family may have more interaction with his parents than the first born from another family. For these reasons, birth order can be only a rough indicator of adult involvement. Used rigidly to evaluate one individual's early family experience, it can easily mislead. But in detecting trends in large groups it may be of considerable value.

My initial interest in birth order stemmed from a study of boys with at least average intelligence who had been referred to the Institute for Juvenile Research primarily because of learning problems. Comparison of a hundred such boys with another hundred boys who had been referred to the Institute for problems other than learning (e.g., anxiety, social maladjustment) revealed birth order to be one of significantly differentiating variables. More specifically, it was found that first born boys were apparently less susceptible to learning problems than were last born boys.[7]

I then reviewed the literature on birth order and discovered that there has been a low keyed debate over its value as a concept or an indicator. Birth order, it appears, had been picked up by investigators, followed enthusiastically for a time, and then, for a variety of reasons, dropped. Because there may be still some resistance to a serious consideration of birth order, it may be helpful if I briefly indicate why I believe there have not been good and sufficient reasons for dropping the topic.

Particularly is this the case in the depth psychologies. Its formal neglect by psychoanalysis is most likely due to its being considered an Adlerian concept. Yet, five years after Adler broke away, Freud could still write in his *Introductory Lectures:*

> "Forced into second place by the birth of another child and for the first time also entirely parted from the mother, the child finds it very hard to forgive her for this exclusion of him. . . . a child's position in the sequence of brothers and sisters is of very great significance for the course of his later life, a factor to be considered in every biography."[5]

Another indication that psychoanalysts are attentive to birth order (despite the absence of formal papers on the subject), can be seen in Ernest Jones' biography of Freud. He wrote:

> "We can gather that (Freud) appears to have been a normal sturdy child and we can note the few features that distinguish his circumstances from those of the average run of children. They are few but important. He was the eldest child, at least of his mother, and for a time therefore the center of what may be called the inner family. This is in itself a fact of significance, since an eldest child differs, for better or worse from other children. . . . It may give such a child a special sense of importance and responsibility."[8]

In the more precise behavioral sciences, difficulties in replication appear to be the main reason for flagging enthusiasm. These difficulties inspired such discouraging literature reviews as Murphy, Murphy and Newcomb made some thirty years ago in their experimental social psychology in which they rather flatly stated that the results of studies attempting to find a relationship between birth order and intelligence, personality traits, etc., were inconsistent and contradictory.[13] While replicatory difficulties are understandably often a deterrent to further investigation, they can also constitute a challenge. This kind of challenge may be seen in the review made by H. Jones in the Manual of Child Psychology. He wrote:

> "In an earlier critique (Jones, 1933) of research in this field, evidence has been given that when. . . methodological difficulties are properly controlled. . . . no birth order differences in intelligence occur in normal samples . . . atypical results, however, have been encountered in certain highly selected samples . . . Studies of gifted children by Terman (1925) and of eminent men by Ellis, Cattell, and Huntington has shown a distribution of birth order different from chance expectation and strongly favoring the first born. . . . No satisfactory explanation of this finding has been given."[10]

The foregoing considerations and in particular the review by Jones persuaded me to take seriously the possible value of birth order and to endeavor to account for the results in the learning problem study. With the question in mind "Were first borns indeed more gifted or creative?" and following the lead of Ellis, Cattell and Huntington, I chose for my subjects a large number of men who had been pre-eminent in Western Civilization since 1600. Because biographical information was frequently lacking as to the exact birth position these eminent men had in relation to older and younger brothers and sisters, my compromise solution was to center in whether the eminent man was the first or only surviving male offspring of the mother or whether he was a later male offspring. Of the thousand eminent men investigated, there was an almost equal division between first sons and later sons.

Needless to say, the first direction of the inquiry ran almost immediately into an insuperable obstacle. There was no way of measuring whether one eminent man was more creative than another. And one would be hard put even to estimate whether first son Einstein was more gifted than later son Darwin or first son Jefferson than later son Hamilton. Consequently the assumption was made that the superior intellects of these men were of equal quantity and that any discoverable differences between first and later sons would reside in the directions their creativities took. The difficulties involved in evaluating these directions were only slightly less formidable. But by recourse to some original writings and especially to the authorities in the various fields, it was possible to discern what appeared to be polar divergences in the thinking, feeling and behavior of eminent first and later sons.

This does not mean that every first son thought similarly to every other first son and differently from every later son. It means rather that on certain issues first sons appeared to be clustered toward one side of an issue, later sons toward the other. Examples of this clustering will be given later.

In summary, then, the study of eminently creative men was completely unenlightening as to any relationship between birth order and the quantity of creative intelligence. Quite apart from birth order, it was equally unenlightening as to the baffling question of what makes a man highly creative. But the study did persuade me that, given creativeness, there appeared to be some kind of relationship between birth order and style of creativity.[8]

It would be premature at this point to spell out what this relationship is. Without the background of evidence and theory which shall shortly be given, the reader would be unprepared for the statement that first sons are prone toward seeing things as connected, whereas later sons are prone toward seeing things as disconnected. Yet it may be helpful for purposes of orientation to state that the basic hypothesis to be offered here is essentially that birth order is a rough indicator of a proneness toward connectedness or disconnectedness.

It is in the area of language that the findings are particularly suggestive. The evidence—impressionistic as well as experimental—appears to indicate what may be the repercussions in the child's intellectual development if he has been exposed to more parental involvement (as *tends* to happen with first borns or first sons) or to a lesser involvement (as *tends* to happen with later sons). It indicates that the first born (or first son) has a special proclivity for language, a sensitivity to its structure and meaningful use. This repercussion of parental involvement is of course not surprising. The first born's verbal skills would appear to stem from the fact that he tends to be talked to more frequently by the parent, to be more often a participant on an adult-child dialogue. In this dialogue, he would have to understand what is said to him and he would have to respond in a fashion meaningful to the parent. In short, in verbal intercourse with the parent, the first born would have to adopt the viewpoint of the parent. He would more quickly leave the stage of monologue which Piaget has described as characteristic of child language even beyond entry to school.

Indications of this language proclivity were seen in the eminent men. There was, for example, a noticeable over-representation of first sons among the major poets—Shakespeare, Goethe and Pushkin being prime examples. There were also signs among first sons of a studious absorption in words and meaningful language—illustrative creative products being Samuel Johnson's *Dictionary of the English Language,* Roget's *Thesaurus,* H. L. Mencken's *The American Language* and George Bernard Shaw's devising of a new alphabet. The flavor of this kind of absorption can be seen in what first son Edmund Wilson wrote in a recent *New Yorker* article entitled "My Fifty Years with Dictionaries and Grammar."

> "I have always been greedy for words. . . . I love Elizabethan plays, dictionaries of slang and argot. . . I suppose that it is partly that I like to recapture the excitement I felt in childhood when attacking for the first time on my own a greenish-gray primer, I found that the sentences ran on with meaning."[18]

The principal pertinent experimental evidence can be found in the work of Helen Koch whose investigation of five and six year olds from two-children families is perhaps the most carefully controlled experimental study on birth order. She found in relation to understandable speech that "first born children articulate consistently better than second born children."[11] She also found that those first born sons who were two to five years older than their siblings scored significantly higher on the verbal subtest of the SRA primary mental abilities test. This verbal subtest according to Koch "is heavily saturated with Thurstone's factor V (verbal meaning)."[12] I might add that in a child guidance clinic population, I have found that first born boys scored significantly higher on the verbal section of the Wechsler Intelligence Scale for Children than do later born boys.

The basic issue stemming from language and the necessity for meaningful connected dialogue centers around how much and how soon the child adopts the adult viewpoint. The first born, for reasons described, must adopt this viewpoint in order to engage in dialogue with the parent. In so doing he makes a synthesis between his own world and the adult world. Additionally, he receives the first full impact of the parent's indoctrination of prohibitions and values. The first born or first son is more likely then to be connected with the past, with the older generation and its traditional values. He stands midway as a connecting link between the older generation and his own. The later born or later son is less involved with the older generation or the past. His companions are more often his older siblings of the same generation or his peers. He can remain longer in the stage of expressive monologue and can address himself to the tasks of the present, to those of his own generation.

Considerable evidence exists attesting to the first born's link to the preceding generation, to the bond with the parents which explains his adopting the adult viewpoint. MacArthur, a social anthropologist investigated personality characteristics of first born and second born children (from families of two children) in two generations—the Harvard graduate and his children. Among first child traits, 'adult-oriented' and 'serious' reach statistical significance for combined generations.[13] Seriousness, an adult characteristic and a repercussion of sharing the responsibilities of the preceding generation was found by me to be significantly present in the first born boys of the learning problem study.[7] Another indication of a connection to the past is a conservative tendency to maintain the status quo, to resist a change—a tendency discovered by Hall and Berger to be significantly present in first born children.[6]

A basic way of describing these differences is to say that the first born, due to his experience in meaningful dialogue and his link with the past generation, is prone to see things in connection, to try to make connections by relating one thing to another, he is, so to speak, *connectedness prone.* On the other hand, the later born, due to his experiences in premeaningful language and his less intense link with the past, is prone to see things as not necessarily connected, in fact actually to disconnect the part from the interrelated whole; he is, so to speak, disconnectedness prone. Both mental activities—connecting the parts in a whole and disconnecting the part from the whole—are quite necessary for the optimum exercise of the intellect. But, although superior intellects employ both activities, frequently—and this is the point of the communication—one activity characteristically predominates over the other.

Thus, intellectual styles can emerge from variation in parental involvement—a proneness in first borns toward connectedness, toward a synthesis of the whole—in later borns toward connectedness, toward an analysis of the isolated part. The former would tend on seeing ten fence posts to assume a continuous connection and call the assemblage a fence; the latter would tend to regard each fence post as disconnectedly discrete and to say that there are just ten fence posts, each one of which needed accurate description. These different styles correspond to the two kinds of intellect described by Pascal—the intellect which can entertain many interrelating premises without confusing them and the precise intellect which accurately probes the part.[15]

Evidence supporting the foregoing trends is perhaps not as clear cut as was the case in language proclivity and the tie to the past. Yet, in the writer's view, there are certain trends which do suggest that the first born's creative style is manifested in synthesizing connectedness, the later born's style in analyzing disconnectedness. There is, for example, a list of some twenty eminent mathematicians given by E. T. Bell and divided by him into those who made especial contribution in the interconnected 'continuous' aspects of mathematics (e.g., calculus) and those who made particular contributions in the disconnected 'discrete' aspects (e.g. algebra).[3] First sons predominated in the 'continuous' category, later sons in the 'discrete' category.

A first son—later-son polarity can also be detected among those philosophers and scientists who have theorized about how much connected order there is in the real world. The most outstanding philosophic advocate of disconnectedness is later son David Hume who was skeptical not only about causal connections but also about the existence of a continuous self. Earlier, another later son, Francis Bacon, in his argument for a scientific method based on accurate particulars rather than on synthetic generalities stated that "the human understanding is of its own nature prone to suppose the existence of more order and regularity than it finds."[2]

It is well-known that despite Bacon's strictures, scientific discoveries continue to be made by generalization and speculation as well as by analytic methods. And the great generalizers, the great finders of regularity and connected order, have tended to be first sons such as Galileo, Newton, and Einstein. Einstein's connectedness tendency can be seen in his statement. "It is the aim of science to establish general rules which determine the reciprocal connection of objects and events in time and space."[4] An aversion to disconnectedness, is also evident in Einstein's comment regarding indeterminism in physics that God does not play at dice.

As might be expected, experimental evidence concerning such complex intellectual differences is not easily forthcoming from a study of normal samples. However, an investigation by Morris Stein of creativity in a highly selected sample does have considerable pertinence to what has been just discussed.[18] Stein took a large group of intellectually superior chemists and endeavored to separate them into two groups on the basis of their creative ability. Utilizing the evaluations of the particular chemist's superiors, peers, and subordinates, he was able to set up two groups: the creative and the not-as-creative. He could find no relationship between birth order and creativity as so defined. However, in response to the present writer's query, Stein kindly went over his data and noted that there were significant differences between first borns and later borns as to their score on the Miller Analogy Test. On this test, used to ascertain abstract verbal ability and the facility in making connective analogies, first borns scored significantly higher than later borns. Stein's findings appear to support not only the verbal ability and connectedness trends in first borns but also the general observation that birth order has no discernible relationship to quantity of creativity.

Discussion

There are two matters—both with some clinical implications—which invite discussion. The first touches on the scholastic difficulties associated with cognitive styles. With the increasing emphasis on education, cognitive styles have assumed an ever-increasing importance. A maladaptive style—whether it occurs in elementary school, high school or college—can result in an educational casualty which often seeks counselling or psychiatric help. The first born males, according to the learning problem study on boys, are less apt to be educational casualties in grade and high school. And from Schacter's recent study in which he demonstrated the over-representation of first borns in college and especially in post-graduate schools, it appears that their comparative immunity extends quite beyond high school.[19]

The question still remains as to why the later borns, possessing equal intelligence, do less well as a group in the educational system. That they are less serious than first borns undoubtedly plays some etiological part. Another possible explanation is suggested by the present study, i.e., that the later borns are somewhat penalized for not having the verbal or language proclivity type of intelligence which is so welcomed by educators. This type of intelligence is positively rewarded in the school system and is tested for in most college entrance examinations. The Miller Analogy Test on which Stein's later born chemists scored lower than did the first born chemists is frequently used as a screening test in post graduate levels.

The other matter with clinical implications is considerably more speculative. It raises the question of whether differing theories about human nature may not in part be reflective of differing creative styles. If it is true that first son mathematicians tend to theorize about the 'continuous' aspects of mathematics, whereas later sons tend to theorize about the 'discrete' aspects it is not unlikely that first son and later son psychologists would tend to differ as to what they saw and theorized about in human nature. The likelihood of this occurring is suggested by a comparison of first son Freud and later son Adler. We have already noted that they perceived somewhat different aspects of birth order, Freud sensitive to the experience of being "forced into second place by the birth of another child" and finding it difficult to forgive the mother; Adler sensitive to the competitive power struggle between siblings.

Equally to the point in this comparison are the attitudes of Freud and Adler to the significance of the past. Freud displayed a characteristic first son connectedness to the past in his concept of the tradition laden superego and in his therapeutic method which stressed the analysis of feelings transferred from early family figures. Adler, on the other hand, showed a characteristic later son disconnectedness from the past in his therapeutic method which stressed an active coping with present reality and an adjustment to current social norms. The only depth psychologist who exceeded Freud in connectedness to the past is another first son, Jung, whose theory of the racial unconscious represents the ultimate in the deterministic significance of previous events.

The probability emerges that theories about human nature are inevitably less than universal inasmuch as the theorizer, by virtue of his particular creative style, is more drawn to perceive certain aspects of the complex psychological process rather than other aspects. The question raised then is not merely academic. In psychiatric therapy, a rigid adherence to a conceptual model overemphasizing (or seeming to overemphasize) the past or the present can easily misfire if the patient's problems are not strictly confinable to the 'then' or to the 'now.' A strong case can be made thus for a discriminating, eclectic use of various conceptual models of human nature.

The question is also not academic if the therapeutic role is extended beyond those who deal with individuals to those in political life who have to deal with the wishes, aspirations, resentments and fears of large populations. The entire developmental direction of a nation can depend on whose theory of human nature prevails. The most celebrated instance of this was of course in the founding of the United States when Thomas Jefferson and Alexander Hamilton were opposed, among other things, regarding their views on the nature of Man. Their views differed as to whether or not Man has a strong internalized regulating memento of the past, the conscience or superego. First son Jefferson whose intellectual affiliation was with first son John Locke, felt that Man did have such an indwelling mechanism and did not require much external policing. Later son Hamilton whose intellectual affiliation was with later son Thomas Hobbes doubted the significant strength of internalized past-given morality and believed that the State must be ever watchful as the external policeman.

Two hundred years later, the polarity about the significant past became evident when it appeared for a while that the late President Kennedy and Senator Goldwater may be contestants for the presidency; later son Kennedy emphasizing the present and future, first son Goldwater the past values and the status quo.

Apart from these speculations, the trends in creative styles do have practical implications regarding the management of creative talents. It would seem that any successful management must be predicated on the assumption that civilization has benefitted from the division of intellectual labor between first sons and later sons, between the connectedness-prone synthesizers and the disconnectedness-prone analyzers. The obvious corollary is that differences in intellectual styles should be appreciated and respected. Our concern in this democratic age with the pressing social problems of raising the educational sights of the majority of society has perhaps temporarily obscured for us the problems confronting that much smaller portion of society that constitutes our potential creative leadership. Appreciative respect is often a vital requirement for those who use their talents constructively rather than destructively. If one is stouthearted and determined as the great men are, an impatient lack of understanding may have no deleterious effect; in fact it may be a spur. But, if one is subject to discouragement, one's creative potential may undergo considerable attrition. And this is a situation that no society which aspires to be vigorous can well afford.

REFERENCES

1. Adler, A.: The Practice and Theory of Individual Psychology New York, Harcourt, Brace and Company, 1927.

2. Bacon, F.: Novum organum. *In* Philosophers of Science, S. Cummins and R. N. Linscott, eds., New York, Random House, 1947, p. 91.
3. Bell, E. T.: Men of Mathematics. New York, Simon and Schuster, 1937.
4. Einstein, A.: Out of My Later Years. New York, Philosophical Library, 1950, p. 27.
5. Freud, S.: A General Introduction to Psychoanalysis. New York, Liveright, 1933, pp. 289-290.
6. Hall, E., and Barger, B.: Attitudinal structures of older and younger siblings. Meeting of the American Psychological Association, August 29, 1963.
7. Harris, I.: Emotional Blocks to Learning. New York, The Free Press of Glencoe, 1961.
8. ——: The Promised Seed. New York, The Free Press of Glencoe, 1964.
9. Jones, E.: The Life and Work of Sigmund Freud, vol. I. New York, Basic Books, 1953, pp. 13-14.
10. ——: The environment and mental development. *In* Manual of Child Psychology, L. Carmichael, ed. New York, J. Wiley and Sons, 1946, p. 628.
11. Koch, H.: Sibling influence on children's speech. J. Speech & Hearing Disorders 21:322-28, 1956.
12. ——: The relation of 'primary mental abilities' in five and six year olds to sex of child and characteristics of his sibling. Child Development 25:219, 1954.
13. MacArthur, C.: Personalities of first and second children. Psychiatry 19:47-54, 1956.
14. Murphy, G., Murphy, L., and Newcomb, T.: Experimental Social Psychology. New York, Harper and Brothers, 1937.
15. Pascal, B.: Pensées in the Speculative Philosophers. New York, Random House, 1947, p. 209.
16. Schacter, S.: The Psychology of Affiliation. Stanford, Stanford University Press, 1959.
17. ——: Birth order, eminence and higher education Am. Sociol. Rev. 28:757-767, 1963.
18. Stein, M., and Heinze, S.: Creativity and the Individual. New York, The Free Press of Glencoe, 1960.
19. Wilson, E.: My fifty years with dictionaries and grammar. New Yorker Magazine April 20, p. 165, 1963.

Discussion by Raymond Sobel, M.D.

FOR MANY YEARS, birth order has been unfashionable as a concept useful to the psychiatrist or psychologist except when, as Dr. Harris has shown, it bears on "unimportant" items such as sibling rivalry, the inheritance of wealth, and the parent-child relationship! Why then has it been neglected in any considered way? Parents have been telling us child psychiatrists about the differences in raising first and second children for some time and with good reason. It is only recently that some of us are asking our supervisors about the patients' ordinal position as the result of our interest in family diagnosis and treatment, and others even get them to speculate as to the meaning of being a first as opposed to a later child. Parents sometimes comment that the first child should be like a disposable Lily cup which the parent could use and throw away or, as one of my colleagues has put it, the first child should be like the first pancake. There is no question that being a first born is a matter of considerable importance for the future of the child and often his family and society at large. Why then the neglect of the subject?

The facultative psychologists of the twenties and thirties who were searching for single answers to complex behavior and failing to find it in ordinal position, abandoned its use as a concept since it provided no high correlation with the manifest behavior of the subject or patient. They dropped serious consideration of this and other factors such as parental occupation or social class because of a tendency toward oversimplification, occasioned by the lack of or unfamiliarity with the more powerful statistical concepts and tools which the contemporary researcher and theoretician has available to him. Multivariate analysis "discovers" variables which have no overt single behavioral manifestation but which emerge from communalities between diverse responses or traits. Another reason for the neglect may reside in the disinclination of the psychoanalytic profession to treat its data statistically, a tendency which has not been entirely worked through even now. This again may be in the nature of a reaction formation against the stimulus-response psychology of the twenties and thirties or may be the result of the lack of training of psychiatrists in research design and analysis of results.

I was a bit unhappy in reading Dr. Harris' paper to find that he did not report the degree of relationship between creative style and birth order and I would like to know how strong the trends were toward being first born and manifesting synthesizing connectedness, for as Gertrude Stein has said, "A difference to be a difference has to make a difference."

However, it would seem more important for us to look at Dr. Harris' basic hypothesis and the evidence which he adduces for it. He suggests that "birth order can be fruitfully employed as a rough indicator of basic and divergent psychological tendencies and that there is a major effect of these tendencies on the manner in which the creative individual perceives the world; not only the world of things, but also the world of ideas and of relationships." It is a deceptively simple hypothesis with perfectly obvious logical consequences: Being a first born means closer relatedness to parents. This brings in its train a more meaningfully connected dialogue with significant adults which in turn provides for more opportunity for identification and introjection of the parents as well as for increased traditional acculturation.

As the result of greater linguistic consensus between the first born and the parent, an intellectual style develops characterized by connectedness and endowing perceptual processes. This sequence of explanation is lucid, orderly and logical. It fits the principle generalized by Whorf[1] as the doctrine of linguistic determinism in that the structure of perception and thought is determined by the structure of the language one learns. However, just because proneness to connectedness pertains to first borns, it is not a necessary corollary that a proneness to disconnectedness will exist in nonfirst borns, i.e., later borns. I do not doubt that connectedness is a basic psychological tendency or cognitive style nor is there any reason to doubt that there is a corresponding tendency toward disconnectedness, although I would prefer separateness as a term. It is not certain, however, that disconnectedness is linked with the later born condition even though it is a polar attribute. Dr. Harris has not established the latter as firmly with the data at hand.

Dr. Harris has revived an interest in typology which also had its heyday in the 1930's but which began with Galen in his humoral typology of the choleric—phlegmatic, sanguinic—melancholic axes. Early in the 1900's Jordan,[2] Gross,[3] and Jung[4] postulated similar typologies which included variations on what we now call cognitive style. This approach has reached its apogee in the work of Eysenck and Cattell[5] whose multivariate correlational analyses have been used to establish the validity of perceptual and psychological types or their less inclusive representations in behavior, namely traits and habitual responses.[6] Although Eysenck is less than popular these days among the psychoanalytic psychotherapists, he and Cattell, more than any others, have analyzed typology as a basic approach to personality, making them strange bedfellows with Dr. Harris. In his chapter on personality traits and measurement, Eysenck reviews[7] the literature on *Spaltungs fahigkeit* or "ability to dissociate . . . to form separate and partial groupings within a single act of consciousness; the ability to dissect complex material into its constituent parts." He then reports a large number of experimental methods which would seem to test this ability but which when subjected to factor analysis show no relationship to each other. This is germane to Dr. Harris' thesis, since what we feel to be evidence of a trait in a variety of criterion performances may fail to show up when more rigorous statistical treatment than inspection of the data is employed. In Harris' report of Stein's data on creative and not so creative chemists it could be that the differences on the Miller Analogy Test were due to a factor other than connectedness, possibly Spearman's "g" or Thurstone's second-order factor of intellectual ability. Were the more creative chemists brighter than the not so creative? There seems to be no control for this variable. If Dr. Harris is going to use this experimental approach, I believe he is obliged to control his data for factors other than the one for which he is searching since it may not be the proneness to connectedness or disconnectedness that varied with the order of birth but rather another factor entirely. This has proved to be the case in many recent studies in psychopathology in which, for example, the independent variable turns out to be social class rather than a dominating mother or a passive father.

The foregoing should not discourage us from pursuing Dr. Harris' postulation of a connectedness-disconnectedness axis. There is considerable evidence in the literature on personal constructs and cognitive functioning that this tendency is present but masquerades under different names. I should like to review a few of them for you since they tend not only to give depth to the concept but also provide clear cut experimental methods by which the connectedness-disconnectedness tendency can be measured and validated. If this is the case, then we have the opportunity to correlate the behavior of the individual on these tests with his ordinal position and thus confirm or invalidate this particular typological scheme.

Asch has designed a group-interaction procedure in which the individual is placed under pressure by his peers to agree with a trumped up and egregiously erroneous group opinion. Basically it consists of having a group of 8-16 persons judge which of three lines is the same length as a standard, with each subject in turn announcing his choice in a loud voice. Actually, all the subjects except one are "ringers," hired phonies, and it is rigged so that the "real" subject responds toward the end of the series.[8] Two types emerge, the Yielders and the Independents. When these two groups are asked to describe themselves on the Gough Adjective Check List which consists of 279 commonly used personal adjectives, e.g. artistic, forgetful, kind, etc., the Yielders tend to rate themselves more like the connected synthetic type described by Harris where the Independents tend to be more like the disconnected analytic type.[9]

Barron, using MMPI scales and attitude questionnaires to derive a typology with polar extremes of simplicity and complexity finds that preference for simplicity is significantly associated with social conformity, respect for custom and ceremony, friendliness toward tradition, etc., the qualities which Dr. Harris associates with Jefferson and Locke, or synthetic thinking. On the other hand, Barron reports that complexity goes along with unconventionality, political radicalism, high valuation of creativity and a liking for change, perhaps more Hamiltonian or analytic in style. Barron suggests that the types of perceptual preference he observes "are related basically to a choice of *what to attend to* in the complex of phenomena that make up the world we experience; for the world *is* both stable and unstable, predictable and unpredictable, ordered and chaotic. To see it predominantly as one or the other is a form of perceptual decision. . . . Either of these alternative perceptual decisions may be associated with a high degree of personal effectiveness." Barron speculates on the effect of these opposing tendencies on creative style much in the manner of Dr. Harris.[10]

Witkin and his associates approach the connectedness-disconnectedness theme with a most novel methodology.[11] Using a variety of ingenious apparatuses, they have discovered that people differ in the way they orient themselves in space. Further, they found the way in which the individual does this to be an expression of a more general preferred mode of perceiving which in turn is linked to a broad and varied array of psychological functioning. One of these attributes which would correspond to Harris' proneness to connectedness is the "field dependent" way of perceiving. This kind of orientation is illustrated by the rod and frame test in which the subject sits in complete darkness, facing a luminous rod surrounded by a luminous frame. Rod and frame can be independently tilted, to one side or the other; the subject sees them first in tilted positions. Then while the frame remains tilted, he moves the rod (through his directions to the experimenter) until it appears to him that it is vertical. As you would imagine, some subjects see the vertical as independent of the frame tilt, in reference to their own body position. They bring the rod to the true upright, determining the perpendicular independently of the field context or the frame. Others are field dependent, and find it difficult to overcome the influence of the surrounding field, i.e., to separate the item (the rod) from its context (the frame). Witkin has found that people tend to be field dependent and field independent in a continuum rather than falling into two distinct categories. Now here, it seems to me, is an ideal method of validating Dr. Harris' hypothesis on a control population. If he is right about conceptual style, then first borns should be significantly more field dependent on the rod and frame test than later borns. Certainly the field dependent people fulfill his other criterion in that they have higher verbal than performance scores on standard tests of intelligence. Although he has not explored this specifically, Witkin's reports suggest to me another experimental method which might be applied to Harris' data, namely the ratio of whole (W) to detail (d) responses on the Rorschach. To the best of my knowledge this has not been studied in relation to ordinal position but it should not be a difficult task. Again, if Harris is right, first borns should perceive more whole responses than later borns and the theory as applied to creativity would be much stronger than Harris' current evidence would support. Lastly, the use of the embedded figures test suggests itself. This is a test which requires the subject to separate—or disconnect—an item from the field in which it is incorporated but which does not involve the use of complicated apparatus such as a tilting room with a tilting chair. These Gottschaldt figures are employed in the same way as some of the Sunday Supplement puzzle contests, the "how many rabbits can you find in the picture?" sort of thing. Since norms have already been established for all of these tests, I would like to see what covariance there would be with birth order and estimation of connectedness as a creative style, particularly since factor analysis has already proved field dependence and field independence to be loaded on different factors. In fact, Witkin has postulated a parallel, or possibly an identical, approach to that of Dr. Harris calling these perceptual styles the "global field approach" and the "analytic field" approach. It would be interesting to test the hypothesis that creative persons identified by Harris to have a synthetic cognitive style would have a global field approach as tested on Witkin's apparatus, on the Rorschach W and D responses and quite possibly on other projective devices which allow quantification of cognitive style.

Another possibility for validation of the correlation between birth order and Harris' typology would lie in the application of the Goldstein Scheere tests to the problem. Here again known synthetic thinkers should score more heavily in abstractness than concreteness and there should be, if the hypotheses hold up, more first borns in this group.

Lastly, I should like to quote from Witkin, "A number of cognitive dimensions, identified by different investigators, may tap the same core of individual functioning. Field dependence, Thurstone's flexibility of closure, Guilford's adaptive flexibility, Phillips' spatial decontextualization and Duncker's functional fixedness and "perceptual organization" on the Wechsler Scale all appear to involve the ability to overcome an embedding context."[12] I might add to this list Rokeach's "closed mind," Holtzman and Gardner's concept of sharpness,[13] Balint's typology of object relationships into Ocnophils and Philobats,[14] Jung's extraversion-introversion and Harris' connectedness. Witkin continues "in a period of extensive research on cognitive styles, it is not surprising that there should be overlap or even identity among the cognitive styles established by different investigators. There is clearly a need for studies aimed at codifying these cognitive styles." I believe that Dr. Harris has made a significant contribution to this effort and that the necessary further documentation should be most productive and rewarding.

REFERENCES

1. Whorf, B. L.: Language, Thought and Reality. Cambridge, Cambridge Technical Press, 1956.
2. Jordan, F.: Character as Seen in Body and Parentage. London, Paul, 1890.
3. Gross, O.: Über Psychopathologische Minderwertig beiten Zupzing. 1909.
4. Jung, C. G.: Psychological Types. New York, Harcourt, Brace, 1923.
5. Eysenck, H. J.: The Structure of Human Personality, 2nd ed. London, Ryerson Press, 1960.
6. *Ibid*, p. 13.
7. *Ibid*, p. 53-58.
8. Asch, S. E.: Social Psychology. New York, Prentice Hall, 1952.
9. Barron, F.: Creativity and Psychological Health. New York, VanNostrand, 1963, pp. 170-180.
10. *Ibid*. pp. 195-196.
11. Witkin, H. A. et. al.: Psychological Differentiation. New York, John Wiley, 1962.
12. *Ibid*, p. 80.
13. Holzman, P. S., and Gardner, R. W.: Leveling, sharpening and memory organization. J. Abnorm. & Soc. Psych. 61:176-180, 1960.
14. Balint, M.: Thrills and Regressions. New York, International University Press, 1959.

PERSONALITY CHARACTERISTICS OF THE POTENTIALLY CREATIVE SCIENTIST

By MORRIS B. PARLOFF, Ph.D.* AND LOIS-ELLIN DATTA, Ph.D.

THE PERSONALITY theorists' interest in "creativity" long antedates the current popular emphasis on this area. Despite this, it must be acknowledged that even our most highly developed personality theories are based primarily on groups characterized by limited rather than optimal functioning. This fact has hampered the development of personality theory and has sharply limited our grasp of concepts such as "mental health," "self-actualization," and "creativity."

An investigator who undertakes research in the field of "creativity" frequently experiences the rather giddy sensation of having left science and joined a literary club. Numerous elegant statements may be found regarding creativity, many of which might be gratifying to the muses of the creative arts, but unfortunately, very few are satisfying to the bemused of science.

If we remain in the idiom of the writer, then we may look upon much of what has been written regarding creativity as directed toward answering the six questions that every journalism student strains to include in the lead of his news stories: who, what, why, when, where, and how.

Those concerned with the *who* issue have undertaken descriptions of individuals and groups on the basis of personality, sociological and biological variables.

The *what* question concerns the qualities of products that are judged by society and its delegates to be deserving of the label creative. This type of evaluation is developed most highly, perhaps, in literature

*Grateful acknowledgment is hereby made of the assistance of Marianne Larson, and Claude Southerly of this laboratory and Dr. Samuel W. Greenhouse, Chief, Section on Theoretical Statistics and Mathematics, NIMH. We are indebted to Dr. Watson Davis and Dorothy Schriver, Director and Associate Director respectively of Science Service Inc. and to Dr. Harold Edgerton, President, Performance Research Inc., for their cooperation in every phase of this study.

and the arts. From creative products the extrapolation is made that the processes by which they were produced must be creative, and that the man in whom these processes churned must certainly be creative.

Those concerned with the *why* of creativity have focussed on the nature and degree of motivations, conscious and unconscious, which are assumed to energize the creative man (e.g., Brill, 1931; Kubie, 1953; Sachs, 1951). Research dealing with *when* includes studies of the interaction of individuals and specific environmental forces (Stein, 1956). These studies also consider the ages and cycles within the individual's life span when he is more and less creative (Lehman, 1953). The issue of *where* has been investigated in terms of ecological and demographic factors (Cattell, J. McK., 1906; Visher, 1947). Other researchers have described the *Zeitgeist* which characterizes various cultural epochs, as well as the characteristics of specific settings which appear to be more or less conducive to creative productivity (Hulbeck, 1945; Merton, 1964). And, finally, the *how* issue concerns the processes which are hypothesized to explain the sequence of events which appear to culminate in a creative act or a creative product (Wallas, 1926).

This study focusses principally on the *who* question, for we are concerned with the personality development of scientists who are designated as creative. Current concepts regarding personality characteristics of creative scientists are derived from studies of eminent well-established scientists. Such characteristics may, therefore, reflect traits and organizations of traits developed in response to recognition of creative behavior but need not have been determinants of creative performance.

The validity of conclusions regarding the early socialization experiences of the creative individual is limited by at least two facts. First, retrospective accounts of eminent mature scientists may reflect idealized views—positive or negative—of their formative experiences. Second, longitudinal studies have been of limited scope. Pioneering studies such as those of Terman have included relatively few gifted children who grew up to be scientists (1954). As a consequence, these studies have shed but little light on the question of what early socialization experiences distinguished the gifted children who became eminent scientists from those who did not.

Recognizing the low incidence of scientists who are ultimately designated as creative, we decided to study a highly selected group of adolescent boys who give promise of being creative in some area of science. We intend to assess their personality development periodically over an extended period of years. Social history information will be sought from these subjects and their parents, teachers, and others.

We assume that data based on retrospective accounts covering periods of fifteen to seventeen years may be somewhat less distorted than accounts which span periods of fifty to sixty years. The accuracy of this assumption will be subjected to test when our samples are periodically reinterviewed.

We believe that such a longitudinal study will provide an opportunity for the increased understanding of the following four problems: (1) the qualities or propensities of an individual which contribute to creative performance; (2) personality changes within the developing scientist; (3) personality features of scientists prior to and subsequent to entrance into particular vocational fields (and functions); and (4) the interaction of personality, cognitive and environmental factors in the successful and less successful scientist.

The design of a longitudinal study requires the selection of subjects who differ in degree of rated potential creativity but are comparable on such relevant variables as intelligence and scientific aptitude. We assume that personality variables are related to creative performance; therefore, groups selected on the basis of different degrees of creative performance should be found to differ on relevant personality characteristics. This study asks two major questions: (1) Is there evidence (statistical and clinical) of personality differences among these groups; and (2) Are the personality characteristics of the potentially most creative group consistent (statistically and clinically) with personality characteristics of known adult creative groups? Positive answers to both questions would encourage us to undertake a full-scale longitudinal study.

SUBJECTS

The sample was selected from male high-school seniors who competed in the 1963 Westinghouse Science Talent Search (STS). This competition is aimed at discovering ". . . boys and girls whose scientific skill, talent, and ability indicate potential creative originality. . . ." (Edgerton & Britt, 1945). Out of the more than 2500 male applicants who successfully completed all entry requirements, we selected the 572 who scored above the eightieth percentile on a science aptitude examination devised by Science Service, Inc. Each of these subjects had also submitted to the Science Talent Search judges a report describing his independently conducted research project. The project was initiated by the student and conducted by him over as long a period as he felt necessary—ranging from a few weeks to three years. The projects were scored for "creativity and potential creativity" by the judges of the STS. We believe that the product criterion is directly relevant as a predictor since it is most comparable to the criterion which is used to assess the contribution of the adult scientist. The standards for judging science projects have been refined over the period of twenty-two years since the competition was initiated. Essentially the criteria included evidence that (1) the applicant recognizes and formulates "novel" relationships; and (2) such formulations are plausible and/or effective. The judges were selected on the basis of their acknowledged eminence in one of the following fields: Chemistry, Mathematics, Biology, Medicine and Physics. Each project was independently rated by a minimum of two judges expert in the area of science concerned. The judges assigned each project a letter grade ranging from A, "unusually original," to E, "very pedestrian." All ratings which differed by more than one grade required a conference of the two judges to determine and resolve the basis of the discrepancy. The final ratings all met the required standard of agreement. On the basis of these ratings the sample was subdivided into three groups to reflect different levels of creative potential. Group I, High Potential Creativity, included students whose average project score ranged from A to B— (N=112); Group II, Moderate Potential Creativity, included project scores ranging from C+ to C— (N=137); and Group III, Low Potential Creativity, ranged from D+ to E— (N=287). Since our selection procedures are not consistent with those of STS, we will refer to our sample as the "Potentially Creative Scientist" sample (PCS).

Instruments

A four-hour battery of personality, attitude, and personal history questionnaires was administered to each of these subjects. Since the students lived in forty-four states and the District of Columbia, it was necessary to administer the tests through the mails. The personality battery, selected on the basis of the reliability of the instruments and the relevance of the factors which they purport to measure, included:

(1) The California Psychological Inventory (Gough, 1957); (2) The 16 P.F. Form B (Cattell, R. B., Saunders & Stice, 1950); (3) The FIRO-B [Fundamental Interpersonal Relationships Orientation-Behavior, (Schutz, 1958)] (4) Ego-Strength Scale (Barron, 1953); and (5) the Welsh Anxiety and Repression Scales (Welsh, 1956). Social history and attitudes material were obtained from an adaptation of Rosenberg's (1963) High School Questionnaire, and an Interpersonal Relationship, Childhood, questionnaire (Shanan & Stein, 1964). This report will not include a full survey of the social background characteristics, but will mention a few findings in order to describe the total PCS sample.

Results

The results reported here are based on a total of 536 of the 572 subjects, or 94 per cent of those originally surveyed. The returns from those classified in Groups I, II and III were 94.9 per cent, 91.5 per cent and 94.4 per cent, respectively. In view of this excellent return, the usual survey problem of selectivity of respondents has been minimized.

1. Age. The students' average age was 17.3 years and there were no differences in age among our groups.

2. Intelligence. On the basis of the Scholastic Aptitude Test Scores (verbal and mathematical), the PCS sample is clearly of high intelligence. The means of each of the three PCS subgroups are above the ninety-ninth percentile for male high-school seniors. No meaningful differences are found among the three groups.

3. Scientific Aptitude. The selection procedure required that each subject score above the eightieth percentile for male subjects. No significant differences were found among the mean scores of each of the three groups.

4. Vocations of Father. More than 53 per cent of the fathers of the total PCS sample are in the professions. This is a significant departure from the 12 per cent found in the general population. There are no significant differences among our groups. About 16 per cent of the fathers in the total PCS sample were classified as scientists as compared to about 3 per cent for Roe's Eminent Scientist group (1953) and 5 per cent for Visher's sample of Starred Scientists (1947).

5. Education of Parents. About 25 per cent of the mothers of subjects in the total PCS sample held college degrees, as contrasted with 5 per cent of mothers in the general population in 1960, according

to the U. S. Bureau of the Census (1963). About 29 per cent of the PCS fathers held bachelor's degrees as compared to 4.7 per cent of white males age forty to fifty-four years. An additional 22 per cent of the fathers held advanced degrees as compared to 4.5 per cent reported in the 1960 census. No significant differences were found among our three PCS subgroups.

6. Socioeconomic Status (SES). An estimate of the father's socioeconomic status based on Hollingshead's Two Factor (education and occupation) Index (Hollingshead & Redlich, 1958) indicates that the total PCS sample comes disproportionately from the higher socioeconomic groups. A total of 59 per cent were classified SES I and II as contrasted with 8 per cent found in a New Haven sample (Hollingshead & Redlich, 1958). Our samples included only 10.4 per cent in SES IV and V in contrast to the 75 per cent found in the New Haven sample. There are no differences among our groups.

7. Intactness of the Family. Of the total PCS sample, only 12 per cent came from homes broken by death, divorce, or separation, compared with the 17.4 per cent found in a survey of 4687 high-school students in the state of New York (Rosenberg, 1963). Our groups are not differentiated on this variable.

Let us now turn to the first question: Do our three PCS subgroups differ significantly on statistical analyses of personality characteristics? For purposes of this report we will consider only personality data obtained from the California Psychological Inventory, which has been well validated on large samples of normal subjects.* The findings

*In interpreting the CPI we assume only that the self-rating constitutes a significant piece of verbal behavior, some of whose correlates the test developers have discovered by empirical means. We make no assumption that the subject is able to describe his own behavior accurately, i.e., that he is aware of what he does, or that if he recognized it that he would tell the examiner about it. Whereas the test situation seems to request an accurate self-rating, the scoring does not assume a valid self-rating to have been given. There is a projective element involved in the subject's interpreting and responding to the verbal stimuli of the questions which is similar to the individual's response to the inkblots. It is the ambiguity of the inkblot which permits the individual to impose his own interpretation of the stimuli; in the same sense the ambiguity of the word and the inaccuracy of human memory provides some sources of discrimination among individuals. Most of the scales of the CPI were developed on the basis of a demonstrated relationship between the behavior being measured and the thirty to forty items of the empirically derived scale. Items selected for inclusion in each scale differentiated among various criterion groups.

based on the other measures are consistent with those to be presented, and merely strengthen and supplement the CPI findings. The CPI has been used widely in other studies of creativity, and therefore comparisons with other groups can be made. The CPI is not sensitive to factors of age differences and does not require rigorous conditions of testing. It has provided valid and useful results when administered under nearly every conceivable condition including informal sessions, take-home plans, mail-back, etc. (Gough, 1957, p. 8). The instrument consists of eighteen scales which survey the individual from the point of view of his social life and everyday behavior.

On the basis of appropriate F tests, significant differences were found to exist among the three PCS subgroups. When a comparison of the eighteen mean scale scores of Groups I and III was undertaken, these groups were found to differ significantly on four scales. Group I was more independent, autonomous, self-reliant and having broad interests (*Achievement via Independence*); more efficient, clear thinking and planful (*Intellectual Efficiency*); more perceptive, resourceful and rebellious toward rules and constraints (*Psychological Mindedness*); and more imaginative and impatient and less like the typical respondent in the pattern of responses to this test (*Communality*) than Group III. With reference to the general norms for the CPI, both Groups I and III scored above the norms on *Achievement via Independence, Intellectual Efficiency* and *Psychological Mindedness* and below the norm on *Communality*.

In comparing the average scale scores of Groups I and II, four significant differences were found. Groups I and II were both below the CPI norms on *Sense of Well-Being, Self Control* and *Good Impression* and above the norms on *Intellectual Efficiency*. Group II, however, was more defensive and showed more constraint and self-doubts (*Sense of Well-Being*); more impulsive, self-centered, excitable and irritable (*Self-Control*); less concerned about making a good impression and more distant in relationships with others (*Good Impression*); and less efficient, clear thinking and planful (*Intellectual Efficiency*) than Group I.

Comparisons of Groups II and III revealed four significant differences. Both groups scored above the CPI norms on *Achievement via Independence, Flexibility,* and *Femininity* and below the norms on *Sense of Well-Being*. Group III, however, was less independent, autonomous and self-reliant (*Achievement via Independence*); less insightful, adventurous, rebellious and assertive (*Flexibility*); more outgoing and ambitious but with a narrower range of interests (*Femininity*); and not as defensive, showing fewer constraints and self-doubts (*Sense of Well-Being*) than Group II.

These findings, which are supported by other results, indicate that the difference between our high and low potentially creative groups are generally linear on such measures of cognitive style as *Intellectual Efficiency*. On measures of psychological adjustment, such as *Well-Being*, the differences are generally nonlinear, Groups I and II being more similar to each other, and Group II being divergently lower. On the bases of these statistical analyses of the separate scales we conclude that our groups are differentiated on the personality measure we have employed.

The second question asks whether the three PCS subgroups differ from each other in a psychologically meaningful fashion. To answer this question we obtained a clinical interpretation of each group's profile. We then submitted five personality profiles to Dr. Harrison Gough, author of the CPI, and asked for a clinical description of each. The profiles bore no identification other than sex and size of sample. The five profiles included our three PCS subgroups, a group of sixty-two male high-school students of high intelligence (median I.Q. 139 on the Terman Concept Mastery Test) and forty-five creative male Research Scientists (mean age of 35.7) selected by their supervisors in an industrial setting on the basis of their demonstrated creativity (Gough, 1961).

The CPI profile of Group I was described as follows: (1) a high level of intellectual ability, i.e., a level consistent with the group's being excellent students with a zest for independent effort and constructive endeavor; (2) a high level of resourcefulness, i.e., adeptness at coping with new and different situations and in evolving effective techniques for dealing with new problems; (3) the capacity for independent work, i.e., the ability to generate internal motivation without need of external props and reinforcements; and (4) a capacity for original and innovative work.

Group II was distinguished for its complexity. In part the report stated that: The Group II individual ". . . will be strong, resolute, and forceful and yet at the same time unsure of himself, doubting, and pessimistic. He values himself and his work and seeks recognition and yet is dubious and even annoyed when it is forthcoming. He welcomes change, uncertainty, and emotional expressiveness, and yet is made anxious by the strength of his own emotionality. He is, in short an individual of talent, complexity, and creative potential."

Group III was described as appearing to be less concerned with individuality and freedom from external coercion, and showing less of the spontaneity which leads to creative expression. "There is less aggressiveness and counteracting inertia and complacency but, likewise, less zest for innovation and departure. The overall implications are, nonetheless, highly favorable. The Group III student should be intelligent, insightful, autonomous, constructively motivated, resourceful, and able to persevere in pursuit of meaningful goals."

According to clinical descriptions, then, all three groups seem to be highly motivated and effectively functioning, but Groups I and II are judged to be more creative than III.

Let us now turn our attention to the final question: Do the personality characteristics which differentiate potentially creative scientists resemble those said to distinguish eminent creative men? In this section we will consider only Group I of our PCS sample, i.e., the group judged to show the highest potential creativity on the basis of its project ratings.

If we find no differences between PCS I and known creative groups, then we would conclude that these characteristics are developed fairly early in the life of the scientist. If differences are found, then the assumption that personality features of creative adults may be a consequence rather than a determinant of creative performance remains tenable although far from established. Great caution must be exercised in interpreting any differences since the bases for such discrepancies are not clear in this study. One of the major sources of variance could be the influence of different degrees of creativity. The results of comparing a noncreative group with a highly creative group, for example, may or may not be like the results obtained in comparing a moderately creative with a highly creative group. There is no assurance that any two groups selected as "more creative" than their less creative control groups will be equally creative.

The principal comparisons to be reported are among the following three groups: PCS Group I, a group of forty-five Research Scientists,*

*These industrial research workers were primarily physicists, mathematicians, and electronic engineers. "Each man was employed as a researcher, and each was engaged in activity where original endeavor could and hopefully would occur. However, in contrast to other Institute projects the search was not for men of renown or outstanding creativity" (Gough, 1961, p. 6).

and a group of forty of the nation's most creative Architects (MacKinnon, 1961).* The Research Scientists included physicists, mathematicians, and electronic engineers evaluated by their peers and supervisors in industry on creativity. The Architects were men nominated by their peers for unusual creativity and eminence in their field. We have chosen these groups for the vocational similarity of one to the unusual creative eminence of the other.

In comparing the PCS I with Research Scientists we found they were alike on only three scales: *Self-Acceptance, Good Impression* and *Femininity*—i.e., they have a high sense of personal worth, low concern with creating a favorable impression, and are hard-headed, ambitious, and broad in interests.

Comparisons of PCS I and Architects revealed similarities on ten of the eighteen scales. Both groups scored high on: *Capacity for Status* (ambitious, self-seeking, broad interests); *Social Presence* (poise, spontaneity, and self-confidence in personal and social interactions); *Self-Acceptance* (high sense of personal worth); *Responsibility* (dependable); *Psychological-Mindedness* (observant, spontaneous, talkative, resourceful); and *Flexibility* (adventurous, confident, humorous, rebellious, assertive). Both groups were low on: *Sense of Well-Being* (are somewhat defensive and self-doubting); *Self-Control* (tend to be aggressive and assertive); and *Communality* (imaginative, restless, impatient). The two groups were about average on the *Achievement via Conformity scale,* (attitudes and behaviors which facilitate achievement where conformance is considered a positive behavior).

In comparing the Research Scientists and the Architects, we were testing, in effect, whether or not there was a core personality common to both of these creative groups independent of their different vocational interests. There were only three scales on which no differences were found. The two eminent groups were both: high on *Dominance* (leadership potential, persistent, self-reliant); high on *Self-Acceptance* (self-confident, outspoken, demanding, aggressive); and low on *Socialization* (somewhat opinionated, stubborn, rebellious). If these three scales are to be viewed as the core of the creative personality, then it is of particular interest that the only scale on which our highest potentially creative group did not differ from either the Research Scientists or the Architects is one of these scales, namely, *Self-Accept-*

*This group was selected to represent the most creative architects in the country. Original data were generously supplied by Dr. Donald W. MacKinnon.

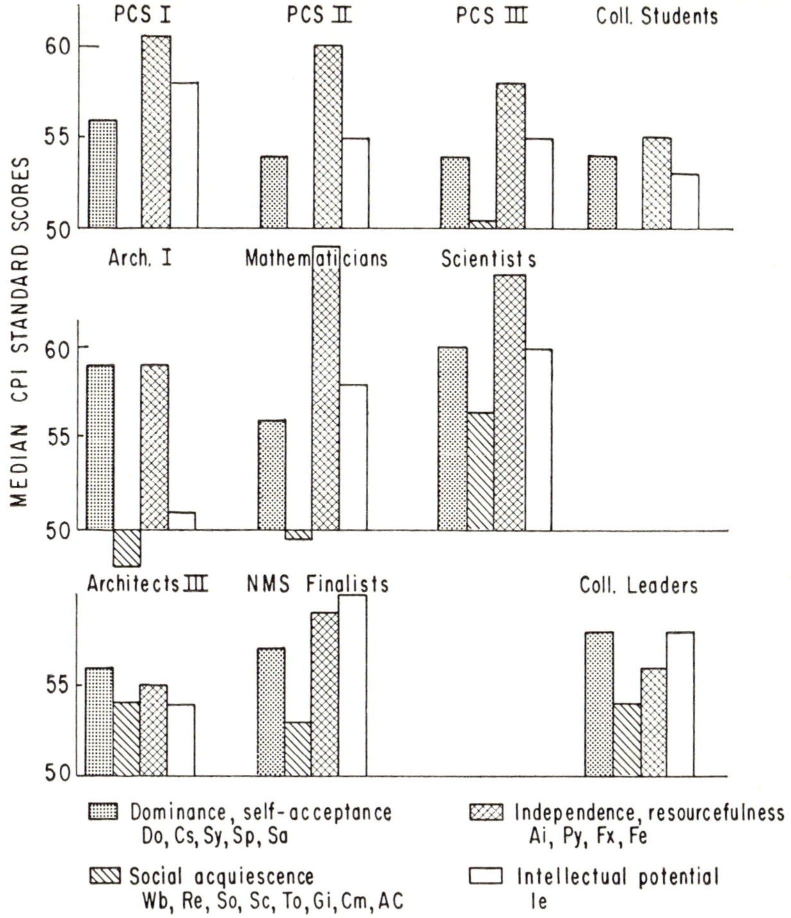

FIG. 1.—Median CPI standard scores for PCS samples and some comparison groups.

ance. It may be noted here that the three PCS groups are each high in *Self-Acceptance,* although PCS Group I tends to be highest.

The fact that PCS I was more similar to Creative Architects than to Research Scientists may be interpreted as support for the assumption that Group I is a potentially creative group since MacKinnon's Architects were more highly selected on creativity than were the Research Scientists.

In order to convey some aspects of the CPI patterns, an experimental index was developed. The eighteen scales were grouped into four categories. The four groupings are as follows: (a) Dominance and Self-Acceptance (Do, Cs, Sy, Sp, and Sa); (b) Social Acquiescence (Wb, Re, So, Sc, To, Gi, Cm, and Ac); (c) Independence and Resourcefulness (Ai, Py, Fx, and Fe); and (d) Intellectual Potential (Ie). The median standard scores on the four dimensions were compared for the three PCS subgroups; three known creative groups: Architects I (MacKinnon, 1961), Research Scientists (Gough, 1961), and Mathematicians (Helson, 1961); and four comparison groups: Architects III (control) (MacKinnon, 1961), National Merit Scholarship Finalists (Holland, 1959), College Students (Gough, 1957), and College Leaders (Johnson & Frandsen, 1962) (fig. 1.). An index which made psychological sense was the relation between Social Acquiescence and Independence and Resourcefulness. It seemed reasonable that a stronger than average need for Social Acquiescence might function as a suppressor factor for Independence and Resourcefulness. The index adds algebraically the median Social Acquiescence score to the median Independence and Resourcefulness score. The distribution of the ten groups on this index differentiates among groups in a seemingly plausible fashion. Mathematicians, Architects I, and PCS Groups I and II are high; Scientists and PCS III fall in the middle of the distribution. College students, college leaders, and Architects III are the three lowest. While this index is admittedly empirical, it does provide some evidence that the groups are not random samples from a general population. Despite very marked differences in criteria of creativity, vocation, effects of sustained recognition and so forth, there are similarities in personality which distinguish the more and less creative samples.

Statistically the evidence regarding the comparability of PCS I to known eminent groups is suggestive but not compelling. Let us now consider two pieces of clinical evidence related to this question: (1) a blind analysis of CPI profiles; and (2) comparison with a composite creative sample. Gough assessed the relative interpersonal adjustment by ranking our three PCS groups, the Research Scientists and a High I.Q. High-School Sample *from most to least adjusted*. The rankings from most to least adjusted were as follows: most adjusted, Research Scientists, then Group I, Group III, Group II, and least adjusted, High I.Q. Group. In ranking creativity the order shifted markedly, with Group II given the highest probability of creativity followed by

Group I, Research Scientists, then Group III, and, in last position, the High I.Q. Group. It appears that from a clinician's point of view the complexity and interpersonal turmoil of Group II suggest potential creativity and that Group III shows the least creative potential of the PCS groups compared.

A final clinical appraisal of Group I was obtained by comparison with the personality characteristics of a composite creative sample based on the findings of many researchers regarding creative groups independent of the particular professions represented.

Despite a great reluctance on the part of investigators to describe the creative man, Stein (1962) has assembled a list of personality features most frequently reported as descriptive of creative individuals. Such features include the following: (1) self-assertive, dominant, self-sufficient; (2) less inhibited, less formal, less conventional; (3) persistent, self-disciplined, high energy output; (4) independent and autonomous; (5) constructively critical; (6) widely informed, wide-ranging interests; (7) open to feelings and emotions; (8) having aesthetic values; (9) low economic values, poor businessman; (10) freer expression of feminine interests and lack of masculine aggressiveness; (11) relatively little interest in interpersonal relationships; and (12) "emotional instability" but capable of using their instability effectively.

On the basis of our findings, PCS I appears to be generally consistent with this profile, allowing for the fact that in some instances our measures differed from those used by other investigators, and that we did not attempt to measure values directly. A major discrepancy is that PCS I fails to show much evidence of emotional instability; however, MacKinnon also failed to find this as a characteristic of creative groups. Emotional instability is, however, found to be characteristic of PCS II. In summary, then, it appears that both statistical and clinical evaluations of our three groups indicate that Groups I and II may be more creative than Group III and that clinically if not statistically there is evidence that our PCS I shares salient personality characteristics with known creative groups.

Discussion

Group I was similar to both Architects and Research Scientists on only one of the 18 CPI scales, i.e., *Self-Acceptance*. This fact, coupled with the finding that *Self-Acceptance* is one of the three scales on which no differences are found between the two eminent creative groups, lends importance to this dimension. Since our three PCS groups do not differ among themselves on *Self-Acceptance,* a question is raised regarding the role of self-acceptance in promoting creative work. If it is assumed that all individuals have the capacity to achieve creatively, then presumably a high level of self-esteem will facilitate this process. If, however, we assume that individuals differ in creative capacity then the interaction of high self-esteem with different creative potentials should lead to different consequences. One group's high self-esteem for example, might be based on the fact that its members meet a high standard of creative performance and another group's self-esteem might be based on an uncritical acceptance of lower quality accomplishments. Marked differences in the level of contributions made by these two groups would be anticipated. Self-esteem in the one group may be conducive to creative performance and in the other group may encourage products ranging from pedestrian to bizarre. In the course of the follow-up study, we will have an opportunity to trace some of the implications of this question. There are very likely a number of personality patterns which are consistent with creative functioning. The study reported here, however, considers only average group profiles and has not attempted to identify possible subgroups. Thus, our findings refer only to propensities of the groups from which the samples are drawn. Any attempt to infer individual psychodynamics or individual motivations or drives from such heterogeneous group data would be like looking for a needle in a haystack.

In the field of creativity the issues of criteria and measurement are fully as serious and as far from solution as the same questions in the field of psychotherapy. In the absence of any standard definition, established criteria, or known valid measures, it is extraordinarily difficult to compare the nominees of one judge of creativity with those of another. As in psychotherapy evaluation, we deal not with absolute but with relative criteria. The problems involved in developing an absolute measure of creativity are no less formidable than those of measuring mental health. Until studies are designed systematically to vary criteria of creativity, or until an absolute criterion is devised, the information represented by the literature in the field of creativity will not be cumulative.

On the basis of these preliminary findings we are encouraged to pursue a longitudinal study to investigate the development of our

PCS groups, some of whose members may ultimately be designated as creative.

REFERENCES

Barron, F.: An ego-strength scale which predicts response to psychotherapy. J. Consult. Psychol. 17:327-333, 1953.

Brill, A. A.: Poetry as an oral outlet. Psychoanalyt. Rev. 18:357-378, 1931.

Cattell, J. McK.: A statistical study of American men of science: the selection of a group of one thousand scientific men. Science, New Series 24:658-665, 1906.

Cattell, R. B., and Drevdahl, J. E.: A comparison of the personality profile (16 P.F.) of eminent researchers with that of eminent teachers and administrators and of the general population. Brit. J. Psychol. 46:248-261, 1955.

Edgerton, H. A., and Britt, S. H.: Science talent in American youth. Science 101:247-248, 1945.

Gough, H.: The California Psychological Inventory Manual. Palo Alto, Consulting Psychologists Press, 1957.

——: Techniques for identifying the creative research scientist. In The Creative Person (D. W. MacKinnon, ed.). Berkeley, University of California, 1961.

Helson, Ravenna: Creativity, sex, and mathematics. *In* The Creative Person, D. W. MacKinnon, ed. Berkeley, University of California, 1961.

Holland, J. L.: The prediction of college grades from the California Psychological Inventory and the Scholastic Aptitude Test. J. Educ. Psychol. 50:135-142, 1959.

Hollingshead, A. B. and Redlich, F. C.: Social Class and Mental Illness: a Community Study. New York, John Wiley & Sons, 1958.

Hulbeck, C. R.: The creative personality. Am. J. Psychoanalyt. 5:49-58, 1945.

Johnson, R. T., and Frandsen, A. N.: The California Psychological Inventory profile of student leaders. Personal. & Guidance J. 41:343-345, 1962.

Kubie, L. S.: Some unsolved problems of the scientific career. Am. Sc. 31:596-613, 1953.

Lehman, H. C.: Age and Achievement. Princeton, Princeton University Press, 1953.

MacKinnon, D. W.: Creativity in architects. *In* The Creative Person, D. W. MacKinnon, ed. Berkeley, University of California, 1961.

Merton, R. K.: Studies in the sociology of science. Paper read at the National Institutes of Health, Bethesda, Maryland, February 5, 1964.

Roe, Anne: A psychological study of eminent psychologists and anthropologists, and a comparison with biological and physical scientists. Psychol. Monogr. 67: Whole No. 352, 1953.

Rosenberg, M.: Society and the adolescent self-image: self-attitudes of 5,000 adolescents. In press.

Sachs, H.: The Creative Unconscious, 2nd ed. Cambridge, Science Art Publishers, 1951.

Schutz, W. C.: FIRO: A Three-Dimensional Theory of Interpersonal Behavior. New York, Rinehart & Company, 1958.

Shanan, J., and Stein, M. I.: The IRC Questionnaire. *In* Explorations in Creativity, M. I. Stein ed. Unpublished manuscript, 1964.

Stein, M. I.: A transactional approach to creativity. *In* The 1955 University of

Utah Research Conference on the Identification of Creative Scientic Talent, C. W. Taylor, ed. Salt Lake City, University of Utah Press, 1956, pp. 171-181.

STEIN, M. I.: Survey of the Psychological Literature in the Area of Creativity with a View Toward Needed Research. New York, New York University, 1962. Cooperative Research Project No. E-5.

TERMAN, L. M.: Scientists and nonscientists in a group of 800 gifted men. Psychol. Monogr. 68. Whole No. 378, 1954.

U. S. BUREAU OF THE CENSUS. U. S. Census of Population: 1960. Detailed Characteristics. U. S. Summary. Final Report PC (1)-ID. Washington, U. S. Printing Office, 1963.

VISHER, S. S.: Scientists starred 1903-1943. American Men of Science: A Study of Colllegiate and Doctoral Training, Distribution, Backgrounds, and Developmental Influences. Baltimore, Johns Hopkins University Press, 1947.

WALLAS, G.: The Art of Thought. New York, Harcourt Brace, 1926.

WELSH, G. S.: Factor dimensions A and R. *In* Basic Readings on the MMPI in Psychology and Medicine, G. S. Welsh, and W. G. Dahlstrom, eds. Minneapolis, University of Minnesota Press, 1956.

CREATIVE PROCESS IN THE ARTS

By ROSE SPIEGEL, M.D.

IN OUR MULTIPLE RESPONSIBILITY as analysts there are three standpoints from which we are concerned with art and artists: First, the art of healing, which is our essential function of service and whose most obvious application is to the psychotherapy of creative people. Second, the search for meaning in general and for understanding of our psychic processes, which does not always directly, obviously, and measurably, result in utility, but is part of the esthetic pleasure from our métier. Third, our role, often unearned, of passing value judgments for society —which is one of *our* larger-than-life attributions.

This responsibility entails self-inquiry in regard to our state of knowledge and of attitude both as private and professional persons. It entails communication with other disciplines, including the arts, and then appraisal of what they have to offer us. And finally it entails our active contribution from professional experience.

There are some large and sobering questions to ask ourselves: How are we equipped to understand artists, their personal difficulties, their artistic endeavors and achievements? What is the relation between psychoanalysis and the arts? What role do the arts play in the inner life of the audience, of the artist? Is there an inner experience specialized to artists? The seven arts (or rather, the eight, if we include film as a separate entity), have their fundamental differences and characteristics; can we find any elemental unity to confirm our speaking of "the" arts? What essentially distinguishes art from science? Discussion of the last question bears on what may characterize the artist in contradistinction to other creative persons.

A working concept of the functioning of art includes as much what the audience experiences, what it demands and rejects, as what the artist gives. This directly bears on the creative demands made on the artist, on his assigned social role and his "image," and on the encounter between the artist and his audience. It directly bears on what we consider his credentials as artist, that is, by what signs we know him—and these have varied strikingly at different times and places. (See following chapter by Harley C. Shands.)

The Universe of Art

Basically, *art is a language, a mode of communication.* I here distinguish *mode* of communication from two other levels of abstraction about communication—one, the *specific instance* of communication through a specific art-product; the other, the global generalization that art *is* communication.

This large generalization is often used as mystique and short-hand for the sense of immediacy and of compression of the meanings which art conveys. It refers not only to the evocation of mood, but to the sense of illumination which accompanies expansion of communication. This sense of illumination, however, can be experienced in other universes of meaning—as in the felt understanding of a scientific principle, the nascent insight in psychoanalysis, the depth recognition of another through love.

The proposition, art is communication, is often misunderstood as the artist's obligation to communicate at any cost, to make himself comprehensible to the audience in its terms. That this is held with reservation by artists themselves was presented by Balcomb Greene in a paper entitled "The Artist's Reluctance to Communicate."[1] Indeed, it is the audience that has been forced to break through the communication-barrier to many major creative innovations of fresh vision —albeit often after the death of the artist.

In *Art as Experience* (1934),[2] John Dewey has gone so far as to say that if the art-product does not have its actual audience, its receiver of the communication, it does not exist. This, of course, is analogous to the philosopher's quasi-paradox that a sound does not exist if there is no one to hear it. Indeed, the tragedy of an artist may lie in his feeling when he lacks the audience that his work—and he the person—does not exist. To help the artist-analysand toward a sense of identity larger than defined by the reception of his art-product and to help him work through his feeling of rejection may be the difficult task of the analyst. In other words, the analyst needs to appreciate the complexity of the reality in the artist's relationship to his audience.

The art-product contains its message and meaning, even if it is like an unread letter or an uninterpreted dream.

The mode of communication which characterizes the arts is best recognized by comparison with that of the sciences. It does not follow that the kind of experience dealt with by the two must be different, but *how* it is dealt with is.

The arts and sciences are two different polarities of symbol and meaning. Cassirer and Langer have familiarized us with the notion that man par excellence is the symbol-making animal—that symbolizing and imaging are built into our organic structure and psychologic functioning; that verbal language and the arts, and both reason and fantasy-processes involve symbolizing activity, though of different form.

Our communication with our inner world and the ambient world outside ourselves, which functions through our primal channels of perceiving, involves both direct experience and its symbolization. And it is on such primal modes—of vision, hearing, touch—that each of the arts builds its imagery and symbolization, transmuted into the imagined world contained in every art-product.

Basic in the vocabulary of the arts is sense-imagery. Imagery may be elaborated into metaphor, or, as in modern visual art forms, be intended to evoke the perceptual experience alone as sufficient.

Where sciences use image and metaphor, say as in a construct of colored balls to represent molecules, it is in the service of intellectual clarification, and any esthetic pleasure is a fringe benefit. The hard sciences, often presented as model for us, strive for linguistically tight denotational meanings in the so-called "lawlike" verbal statements, which pay out "pinned-down" meanings in linear, logical discourse, and which appeal to the explicit structuring of proof. Antithetically, the arts call for sensuous, emotive experience, for intuitional response, with the meanings connotational, implicit, and often highly personal to the audience. The meanings, as in poetry, may shimmer into each other, giving the richness, compression and condensation that Freud described for dreams. What clearly differentiates the use of words by the poetic mind from the scientific, as Jacques Barzun puts it in *The Energies of Art,* are "shifting symbols versus rigid ones."[3] The scientist's private creative experience also may involve imaging and metaphor, but their expression in the public domain demands translation into the scientific language form.

Granting the significant exception of psychoanalysis, the kinds of experience and the objectives dealt with by the arts and by the sciences are different. "There are events or relationships . . . which cannot be adequately represented by scientific symbols. . . ."(Lord Brain) Most strikingly, the arts have their characteristic and essential communication-form for presenting experience, as the sciences have theirs.

All this adds up to the axiomatic requirement that the *artist produce images;* that is his minimal credential.

In the present discussion, the mode of communication by image and metaphor out of which the arts arise is viewed as a broad-based human potential which in the artist reaches heightened expression. But we as analysts suffer from an occupational hazard: we distrust imagery, metaphor, and symbolism. Though we may find a respected niche for the specific art-product, and though we have made our peace with this mode of communication in dreams and fantasy—often in our own metaphor of the 'sleeper's psychosis'—we nonetheless are particularly uneasy with this symbolic mode because of its prominence in schizophrenia.

The specific work of art is a resultant of the artist's personal vision and the world-view of the culture, the expression of the interplay and tension, the discord and harmony between the two visions. Personal vision includes the artist's philosophic position, whether or not he is explicit about it with himself or his audience. It contains his error and his reflections on truth and life. It is colored by his own personality and life-experience in complicated ways about which analysts from Freud on have offered hypotheses. The artist asks us to orient toward his vision and message as valid entities in their own right and not to seek his hidden wound or flaw (except that often he leaves extensive diaries around for posterity.*

To the questions how do we ever break through the molding force of the culture? how is cultural change effected? the part answer is, by means of the interaction between the artist's vision and the cultural view. The creative vision of artist and scientist surely provides illumination in our Platonic cave. As one thinks of the titans of literature: Shakespeare, Goethe, Dostoevsky, Euripedes and Aeschylus, creative vision such as theirs need not discern new values, but rather affirm the eternal verities with fresh vitality or depict the human situation with profound compassion. Or, as in the presentday concern with the "little" person and the antihero, appeal to understanding for lives of quiet desperation.

---

*Another interesting paradox—any field of endeavor, any choice of profession or function in society, leads back to the private person, but somehow the artist evokes most wonder and curiosity.

Though psychoanalysts starting with Freud have honored the grand intuitive psychologists of literature, psychoanalysis itself is a cultural force that has added new dimensions to the arts, granted that some aspirants have used analytic theorems as ersatz solutions to artistic problems.

Art is a dialog between the artist and his culture, as well as with his immediate audience.

Art may light up the passage of seasons in individual life, most poignantly in the artist's own—whether or not intended by him. The paintings, cut-outs and church windows of Matisse, the paintings and ceramics of Rouault, when seen in the time-order of their lives, reveal an evolution, a transcendence from youthful engagement to the vibrant serenity of their old age.

We look for deeper meanings from the artist not only through reportage but revelation.

The art-product contains the artist's intention and meaning, which are not necessarily identical with his vision. His intention is not necessarily communicable in expository verbal language. Indeed, as artists contend, it is *not* to be so communicated (though we nonartists have found that a verbal clue helps—particularly for avant-garde music). His art-product may contain perceptible meanings that his own expository discourse is inadequate to point out. Henry James's exposition on *The Turn of the Screw*[4] does not capture the experience the reader undergoes. This may well be a general case, partly because communication through art involves richer meanings than the linearity of exposition can express, partly because the artist may bear the same relationship to his art-product as the dreamer to his dream, he may have difficulty translating his vision and depth-meanings into explanation. The artist may need to undergo the art-process to ascertain for himself what meaning attracted him to the theme. Shaw wrote in his postscript to *Back to Methusaleh* (1953).[5]

> I let the play write itself. Sometimes I do not see what the play was driving at until quite a long time after I have finished it; and then I may be wrong about it just as any critical third party may.

The artist-writer, says John Ciardi, has the compulsion to write because he "hopes to find himself on his page"; he is in pursuit of a "mysterious act of identity."[5a] But this private pursuit need not be rewarding or interesting to the reader and does not establish why the artist offers his production to an audience.

The artist's claim on his audience is that *it sense his intentions and respond accordingly.* This seemingly simple statement actually permits an open-end and unpredictable range of experience—be it of chance and impermanence in "Happenings," the tenderness of a da Vinci Madonna, Everyman's lurking anxiety as in Edvard Munch's *The Cry,* or of emotion- states alien to us, as in *Lolita.*

Tolstoy has defined the essential purpose of art as the evocation in the audience of the artist's feeling:

> Art is a human activity consisting in this, that one man consciously, by means of certain external signs, hands on to others feeling he has lived through, and that others are infected by these feelings and also experience them.[6]

However, there are pitfalls between the artist's intention and the audience's response. The very newness of his offering may subject it to attack, even from an audience taking the artist's enterprise seriously. Tolstoy himself decried Impressionism and Otto Rank, Expressionism. And indeed, artists decry other artists.

In his classic, *Art as Experience,* Dewey presents the notion that *the capacity for esthetic perception is based on a biologic connection between organism and milieu.* The esthetic perception evidences itself in the sensuous dimension which accompanies all process and experience. He states, "Because experience is the fulfillment of an organism in its struggles and achievements in a world of things, it is art in germ."[6a]

Dewey here links the artist's experience with all human experience, as part of the continuum of normal process. Dewey accounts for artists as a sort of breed specialized for self-realization because of the division of functioning—a division of labor, so to speak—forced on us by the complexity of civilization. That is, the artist experiences for his audience.

However, whether or not we would all be artists in some noble state of nature, I submit that the artist fills a far less vicarious function for us. I suggest that we all need communication with others not only in the direct interpersonal impact but through the encounter in the shared universe of thought and imagination. Further, that many of us experience what is a sense of exhaustion from the exercise of symbolic activity in one's own area and this is relieved, as by replenishment, by turning into the products of the mentalities of others. The artist is a specialist in supplying us with images and meanings, and himself needs to be in a two-way communication with the culture, with the vitality of thought, feeling and imagination of others.[6b]

In short, the audience-experience of the arts depends on a communication-need for the imaginations of others. The arts provide economy of effort in symbolization, the artist having done most of the work for the audience in his 'As If.' The perspective on the audience-need for works of art which Freud offered is within his hypotheses of the unconscious and the instincts, according to which the art-product achieves fantasied gratification of primal wishes for the artist and similarly for the audience. A modification suggested by Hanns Sachs is that what is gratified in both is the sense of guilt; i.e., the reverberating communication between artist and audience is through their mutual sense of guilt. Such explanations may be true on occasion, but do not qualify as a universal proposition.

This is apparent in the literary arts, in which the artist provides another world as escape from one's own, or as perspective on one's reality; he may offer the opportunity for a make-believe identification, or for 'katharsis.' According to Aristotle's conception,[6c] by means of 'katharsis' the audience is purged of the fear and pity aroused in witnessing a work of tragedy—and should then be sobered and wiser without having undergone the vicissitudes of the experience in outer reality.* The role of such functions of the arts is often missed or minimized, as in Dewey's statement:

> In an imperfect society—and no society will ever be perfect—fine art will be to some extent an escape from or an adventitious decoration of the main activities of living.

The escapist theory of the functioning of the arts for the audience, as a kind of opiate for the intellectual, implies a substitutive inner activity for "real" living. However, at the least the arts extend the range of our experience of living. And at various epochs in history the arts have presented a perspective on social reality that, far from escapist, has called for active reorientation of attitude. In the plays of "social significance" of the thirties and the contemporary literature on race problems, in *West Side Story,* Picasso's *Guernica,* Britten's *War Requiem,* the artist, through his emotion, and the imagery and symbolism of his particular art-form, actively addresses society about human error and human values, as soundly as verbal appeals to reason and more eloquently.

Not only his intent and meaning but *the artist's skill and crafts-*

---

*Obviously Freud's usage carries meanings different from Aristotle's.

manship are a dimension of his art, implementing his vision and intent with appropriate grace and authority, substance being shaped into form, the structuring having a logic, a validity of its own. One component of the esthetic experience is the empathizing with excellence of performance, the optimistic version of "There but for the grace of God go I."

In short, we ask of the artist not only his imaging and vision but skill and the organizing mind behind it. In contemporary art it is the artists who challenge the audience and test the traditional tenets of esthetics and values.

Problems of craft have become linked with problems about meaning and have become a central focus, with experiments in their resolution presented as art-products in their own right. Exploring what new meaning and expressiveness unexpected materials are capable of is a current concern that the audience is asked to share. Artists are rebelling against the onus of profundity and are asserting the right to make little rather than grand statements in their art-products, with meanings that are whimsical and mischievous, or that force fresh perception or interpretation on the audience—sometimes, it would seem, simply for the sake of novelty. Are these art? At any rate, as at the turn of the century when the art forms of primitive and exotic cultures provided sought-for new directions, so now the quest for meaning, from Zen to automation, is translated into the language of art.

Art and Psychoanalysis

In considering the universe of art, it would sound as though the artist is larger than life. Certainly he is one of the primal cast of characters on the stage of society. However, if we isolate the creative process from the person, particularly if we ignore the fact that *the artist is a person in a social and historical field,* we compound misunderstanding of his life-situation. Certainly he has a no less complicated relationship to society than the rest of us. He has the problem of earning a living by his art—or otherwise; and of having his art-communication reach some audience through the arbiters and mediators of art and taste, who often form his immediate cultural and social field of interaction. We as analysts need to be sensitive to how these problems, his personality and his creativity interact; how these contribute both to his anxieties and to his creativity; how they challenge his creativity.

What kind of 'message' he 'should' communicate has varied with the social role assigned him through history—whether as integral part of the ruling class of ancient Egypt, courted employee of the medieval church and of Renaissance princes, or as one of the dispossessed in the era of the Industrial Revolution—or, in another line of cultural movement, as the sanctioned image-maker of primitive societies. For the Western artist the social role has become less explicit and he has become more identified with the rebel, the nonconformist, the individualist—at times, not only the voice of 'les misérables', but 'une misérable' himself. In contrast, in Russia the assigned role is explicit with government dictation of form, content, and 'message'; and, for those who accept, security at a price.

The independent and often lonely position of 'art for art's sake' which we tacitly use as our frame of reference is a historical emergence which occurred during the flurry of the European revolutions of the 1830's. By the end of the nineteenth century it was still necessary for Henry James to rebel against the art of fiction being put to service for Victorian morality—its province was to be *all* experience, on which the writer opens his window. The Romantic writers were then turning increasingly to their often troubled inner world of experience as resource for subjects of art, out of which burgeoned new approaches, as Impressionism in painting and Symbolism in poetry.

It is interesting that the late nineteenth century marked the rise of three different but related major cultural phenomena significant for the thinking about creative process in the arts: One, the artist's declaration of independence; second, attacks on the artist in the name of 'degeneration'; third, psychoanalysis. Lombroso had expanded his theory that criminality and psychosis are the expression of constitutional 'degeneration' to include genius—genius as distinguished from talent. This theory Max Nordau further expanded in a quasi-scientific diatribe on artists in his book *Degeneration*,[7] published in 1895.* Meanwhile, Freud took up his scientific championship of those afflicted with so-called diseases of degeneration. His *Studies in Hysteria*,[8] in which incidentally he mentioned his love of literature, appeared a year or so before the Nordau book.

Thus the germinal new trend, of drawing on subjective inner experience as meaningful resource, appeared first in the art of the Romantics and soon after as the *sine qua non* of psychoanalysis.

Up to this point, art and artists have been discussed from perspec-

*An example of his reasoning: It was a defect of vision, inborn and constitutional, that made it necessary for the Impressionist painters to use bright colors and pointillism.

tives not usually thought of as psychoanalytic, but which might well be so included. In particular, the communication aspects of art have been explored. I have suggested that we all need the inflow of imagery and symbols from the psyches and experience of others; that that communication-need makes us all at some time an audience and builds a demand for amassing the symbols and meanings that form our cultural heritage; and that the communication-need is not a variant of Freudian instinctual drives.

Let us now turn to the psychoanalytic approaches. Psychoanalysis has always taken art and the creative process seriously, though often as through a glass seen darkly. After all, the deepest homage to the Greek classics was paid by Freud. Our sober task is to consider what psychoanalysts have contributed to the thinking and knowledge about the creative process.

Freud's views on art and artists derived from a lifelong interest, confined, however, to literature and the plastic arts. They center on two large organizing themes of relationship: to the instinct theory and to the 'economics' of psychic functioning. The writings themselves fall into two categories—abstract formulations of psychodynamics, and beautiful essays focussed on an artist or an art-product. The creative process itself, Freud remarked repeatedly, is a mystery and "the nature of artistic achievement is inaccessible to us psychoanalytically." Actually, Freud's two organizing themes have powerful implications about the creative process.

Freud viewed the art-product as having meanings supporting his theories of infantile sexuality and its wishfulfillment, both for artist and audience. The interpretation of *Hamlet* as a variant of the Oedipus complex appeared for the first time in 1900 as a quite brief comment in The *Interpretation of Dreams*[9] and provided the impetus for Ernest Jones's monograph. In his first paper on art, "Delusions and Dreams" in Jensen's *Gradiva,*[10] written in 1906, both the story and its contained dream are interpreted as though they are a case study of the hero, and indeed the plot and characterization lend themselves readily to the theory of the return of the repressed. He set up the antithesis of poet versus neurotic as the hero's destiny because of the "separation of imagination and intellectual capacity." Freud endeavored unsuccessfully to secure from the author personal information on which to correlate his early life-experience with the novella. In this paper also, Freud stated that novelist and analyst both deal with repressed material, the unconscious, and with the formation of compromises.

In 1908, in "The Poet and Daydreaming," a brief and rather lyrical paper, Freud moved firmly into the position only foreshadowed in the previous paper, namely, that the poetic production is a fantasy concealing the gratification of the writer's childhood wish (and not only his hero's), comparable to everyone's childhood play of make-believe, and that the audience derives from the art-product the equivalent gratification. Freud wrote:

> The true enjoyment of literature proceeds from the release of tension in our minds. . . . [probably from] the writer's putting us in a position in which we can enjoy our own daydreams without reproach or shame. . . . How the writer accomplishes this is his innermost secret.[11]

Also in the year 1908, in his paper "Sexual Morality and Nervousness," Freud added the *concept of sublimation* to the theory of the artist's fantasy-gratification, and drew the broader generalization of the significance for civilization of sublimation.

> The sexual instinct [he wrote] . . . places an extraordinary amount of energy at the disposal of 'cultural' activities. . . . This ability to exchange the originally sexual aim for another which is no longer sexual but is psychically related, is called the capacity for sublimation. . . . [*In*] this ability for displacement . . . lies its value for civilization.[12]

In 1929, in his postwar reflections, "Civilization and Its Discontents," Freud wrote with sad longing about fending off suffering by displacing libido, so that the instinctual aim does not

> come up against frustration from the external world. In this, sublimation of the instincts lends its assistance. One gains the most if one can sufficiently heighten the yield of pleasure from the sources of psychical and intellectual work. When this is so, fate can do but little against one.[13]

In 1911, in the paper "Mental Functioning," which is the forerunner of ego psychology, Freud introduced his second approach to the functioning of the artist, that of psychic 'economics.' Speaking of the overriding of the 'pleasure principle,' characteristic of primary process, by the 'reality principle,' Freud said:

> Art brings about a reconciliation of the two principles in a peculiar way. The artist is originally a man who turns from reality because he cannot come to terms with the demand for the renunciation of instinctual satisfaction as it is first made, and then in phantasy-life allows full play to his erotic and ambitious wishes. But he finds a way of return from his world of phantasy back to reality: with his special gifts he moulds his phantasies into a new kind of reality, and men concede them a justification as valuable reflections of actual life.[14]

After the *Gradiva* and 'Daydreaming' papers, Freud takes as established that the work of art validates his instinct theory and subsequently uses this assumption of validity as the starting point in considering both the work of art and the life of the artist. In *Leonardo da Vinci* (1910) Freud, puzzling and elaborating on the painter's infantile recollection, which was probably fantasied, of a 'vulture' descending on him, interpreted it as an overdetermined sexual symbol. In the painting of the Holy Family with Saint Anne, both the choice of subject matter and the symbol of the vulture* are accounted for by Freud by inferences he draws from the painter's biography. Such construction from biographic data Freud designated as 'pathography.' *Leonardo da Vinci* is remarkable for putting Freudian symbols as links between the painter's life and his work of art. In *Gradiva* the symbolism was applied to the fictional character alone. In Freudian pathography the artist's symbolic communication is taken similarly to the symbolism of dreams, "hidden even from himself," as a mode exclusively in the service of the concealed expression of sexuality. Of course this is a moot point for non-Freudian interpretation of the symbolic mode, which rejects this limitation of meaning.

"Dostoevsky and Parricide" (1928) is another instance of pathography. Freud correlates the theme of the murder of the father in *Brothers Karamazov* with the author's own murderous wishes against his father, accounted for by an unresolved Oedipus complex, a bisexual constitution, and the actually violent character of the father. It is interesting that Freud dissociates Dostoevsky the neurotic man and the "sinner" with whom he has no patience, from the creative artist for whom he has unbounded admiration, and from the conformist to authority, for whom he has contempt. And for the creative artist in this complex personality Freud expressed his bafflement in the classic statement: "Before the problem of the creative artist analysis must, alas, lay down its arms."[16]

Freud made several statements which are eloquent of his admiration for the artist. In "Contributions to the Psychology of Love" (1910), Freud said he was entering the province of the imaginative writer and then contrasted the difference in purpose of the writer and the psychoanalyst.

> Writers . . . have certain qualities which fit them for [the] task; more especially, a sensitiveness of perception to the hidden feelings of others, and

*Which Malraux points out by diagram in the drapery of Mary.[15]

the courage to give voice to their own unconscious minds. But from the point of view of knowledge one circumstance lessens the value of what they have to tell us. Writers are bound to certain conditions; they have to evoke intellectual and aesthetic pleasure as well as reproduce reality unchanged. . . . It is inevitable, therefore, that science should lay hands on the stuff which poets have fashioned so as to give pleasure. . . .[17]

In "Totem and Taboo" (1912), Freud sees in the artist's illusion a remnant of primitive omnipotence of thought and recalls the artist's primitive social role of magician, as in the prehistoric cave-paintings.

Only in one field has the omnipotence of thought been retained in our own civilization, namely in art. In art alone, it still happens that man, consumed by his wishes, produces something similar to the gratification of these wishes, and this playing, thanks to artistic illusion, calls forth effects as if it were something real. We rightly speak of the magic of art and compare the artist with a magician. But this comparison is perhaps more important than it claims to be. Art, which certainly did not begin as art for art's sake, originally served tendencies which today have for the greater part ceased to exist. Among these we may suspect various magic intentions.[18]

In "The Theme of the Three Caskets" (1913), Freud sees in *The Merchant of Venice* and in *King Lear,* the poet's transformation of myth to bring to light again the meaning which is both hidden and revealed in the myth itself. "It is by means of this undoing of the distortion and partial return to the original that the poet achieves his profound effect upon us."[19] This suggests that the poet opposes magic!

In his discussion on *Rebecca West* in "Character Types in Psychoanalytic Works" (1915), Freud presents an excellent concept of the art of the dramatist. "We [rely] on the dramatist to have arranged his conscious dramatic combination in logical accordance with unconscious possibilities."[20]

In *The Moses of Michelangelo* (1914), Freud used his own emotional response to the sculpture as the basis for arriving at the meaning he believes was the intention of the artist and he defended this layman's approach against the critics' method. The aim of the artist, he says, is "to awaken in us the same emotional attitude, the same mental constellation as that which in him produced the impetus to create,"[21] —an interpretation similar to Tolstoy's.

Freud implies that the meanings, the 'message,' in a work of art are not readily communicable in words because they live in the realm of the unconscious and outside of art are accessible only to psychoanalysis—again paying tribute to the insight of the artist.

In "Our Attitudes Towards Death" (1915), Freud presents the view that in the literary arts we the audience seek

> . . . compensation for the impoverishment of life [where people] know how to die and . . . are even capable of killing some one else. . . . We [have] that plurality of lives for which we crave. We die in the person of a given hero . . . and . . . die again with the next hero just as safely.[22]

Freud consistently described the artist as possessing the magical talent which enabled him to dip into his unconscious, to mobilize primary process to gratify both his own primal wishes and those of his audience. He admired the artist's illusion. Though he saw in Dostoevsky neurotic and fallible man as well as creative artist, at no point did he present it as a necessary condition that the artist be neurotic, let alone psychotic. Along with Freud's outspoken formulation that the artist gratifies his primary process wishes through his art is the implicit sense that the artist copes with his neurosis by his art.

A question arises from Freud's approach to the artist and, by implication, to the creative process: Is it a necessary condition that the underlying motivation for a work of art be the gratification of the primary libidinal wishes, first of the artist and then of the audience?

In the Freudian school there then emerged two lines of descent of theory about creativity in the arts, derived from the two major aspects introduced by Freud. One continued in a variety of changes rung on the instinct and libido theories. Greenacre exemplifies the approach closest to Freud's. Bergler and others have extended 'instinct' to include the pregenital range, for instance, orality. According to Brill (1931):

> Poetry is nothing but an oral outlet . . . to express a genuine emotion . . . as it were through a chewing and sucking of nice words and phrases. . . . The omnipotence of thought which originally received its first rude shock when the infant had to cry for the mother's breast but which was finally gratified through breast-sucking, still dominates the poet. . . .[23]

Melanie Klein proposed transformation of guilt about pregenital destructive aggressive impulses as the root of creativity.[24] These and others all are concerned with what Bergler called "*one* part of the artist's creative power—his unconscious drives, defences, and counter-defenses."[25]

The second development of Freud's approach is along the lines of ego psychology and complements the instinctual approach. This is exemplified by Bychowski in "The Metapsychology of Artistic Creation" (1951)[26] and in the monumental work of Ernst Kris.

In his book, *Psychoanalytic Explorations in Art* (1952),[27] Kris considers the creative process in the artist to be the sequential movement of regression from the conscious to the unconscious and back again to the control by the conscious ego, without disruptive threat to the ego. The threat of disruption to the ego was presented as pivotal in the distinction between the psychotic state and artistic inspiration, between psychotic art and creative nonpsychotic art. Furthermore, Kris presented psychological conflict "not only as an unavoidable accessory to personality development, but also . . . as an essential ingredient and incentive."

Kubie, in his book *Neurotic Distortion of the Creative Process* (1958),[28] presents the thesis that creativity is inventiveness and is analogous to free association, which mobilizes the person's information and thought by means of his symbolic functions to the point of their coming into awareness. Kubie makes a point of locating symbolic functions in the preconscious and not in the subconscious. He considers creativity a normal and usual potential and does not discuss sublimation, nor concern himself specifically with art and artists.

An interesting view on the artist was presented by Otto Rank in 1932, some time after his severance from Freud. His voluminous book, *Art and Artist,* is a combination of observation of artists and a dark, rather Wagnerian concept about them, plus a nihilistic prediction. The work is of interest because of his views linking art and neurosis. The drive of the artist is to immortalize himself through his work. The artist suffers from two conflicting pulls, Rank states

> . . . the one which wishes to eternalize itself in artistic creation, the other wants to spend itself in ordinary life—in a word, the mortal man and the immortal soul of man. This universal human conflict, which was resolved through many thousands of years by religion and the art which rested upon this ideology, has become more and more acute and difficult with the growth of individual art, until with modern artists it has taken a form very like that of neurosis.[29]

According to Rank, the artist proves his ability to create by giving his conflict esthetic shape, whereas the neurotic with similar conflicts fails to handle them creatively. This is a formulation reminiscent of Freud's antithesis of poet versus neurotic (1906) and of Freud's theory of sublimation. Rank concluded that:

> The new type of humanity will become possible . . . [only when] artists themselves [achieve] . . . a renunciant attitude towards artistic production. A

> man with creative power . . . will be able to put his creative impulse directly in the service of his own personality.[30]

This concept, by removing in its terms the necessity for the artist to create in art, finally liquidates not only the problem of the artist but the artist himself!

Another theme that runs through psychoanalytic and psychiatric literature and which has important bearing on our attitude toward art and artists concerns the significance attached to the following congery of concepts: primary process; prelogical, archaic, symbolic, or autistic thinking, regression to primary process, paleological. A thorough explication of these terms would belong in the history of psychoanalytic ideas, but a few salient examples at this point will suffice.

Primary process, a psychic system which Freud conceived of as early as 1900,[32] was hypothesized to have the following characteristics: fantasy operations by means of which primal infantile wishes are gratified; immunity to logic and reason and the sense of time as we know them in our objective reality. Thus regression to primary process signifies in this frame of reference the relinquishing of the rule of logic and the reality principle which characterize secondary process. Freud designated primary process as archaic and prelogical in the double reference of individual development and species evolution. He correlated the primary process in dreams with the creativity of artists, including, however, an additional unknown factor, awesome and mysterious. Though he always wrote with high regard for the artist and for fantasy and symbolism, the concept of regression, perhaps contrary to his own view, has often been taken to imply psychologic weakness and disintegration, and a decline to a prior state in the individual's development, all of these being pejorative meanings and values.

'Autistic thinking' was introduced by Bleuler in 1911 as a term extending Freud's concept of primary process thinking to schizophrenia. He wrote:

> The patient thinks in symbols, in analogies, in fragmented concepts, in accidental connections. . . . The autistic thinking is the source of the delusions, of the crude offences against logic and propriety and all the other pathologic symptoms.[33]

Understandably for the clinician this context of autistic thinking has overshadowed Bleuler's brief acknowledgment of "normal autistic thinking" on which is based "a large part of poetry, our tales and myths."

Distrust of 'symbolic thinking,' of a different quality and of greater intensity than in anything written by Freud, emanates from the following typical example from Fenichel's textbook on neuroses, particularly as nowhere in it is there a modifying statement about creativity and art:

> Symbolic thinking is vague, directed by the primary process. It is not only a method of distortion, it is also a part of the primal prelogical thinking.[34]

On the other hand, Kris in his book *Psychoanalytic Exploration in Art* (1952) expresses Freud's spirit and interprets the primary process in an affirmative attitude toward artists.

> The symbols functioning in the primary process are not so much vague and indeterminate as "overdetermined" [which accounts for] the distinctive product of an individual artist.

Kris attributes the difference between the functioning of primary process in creativity and in psychosis to the presence of a strong integrating ego in the former presiding over the regression.

My thesis is that myth, dream and fantasy contain the ore of wisdom and vision. Symbolic expression in imagery and metaphor is a normal and profoundly rational mode of experience and communication of reality, and the meanings of such symbols, in contrast to Freudian symbology, are not categorically sexual. Fromm in *The Forgotten Language* (1951),[35] and Tauber and Green in *Prelogical Experience* (1959)[36] are among those who exemplify this position.[36a,b]

To sum up, by and large in the Freudian views the source of creativity lies in the instincts, which are in conflict with reality, and the process involves 'regression' with ego intactness, whether to the unconscious (Kris) or the preconscious (Kubie). There is difference of opinion as to whether the creative process involves at least neurosis.

The position of the non-Freudian psychoanalysts is that creativity is an inner psychic process in itself, that neurosis is not a necessary condition, and that creativity is an expression of fulfillment. This view is in harmony with the concept of personality as presented by Fromm, Horney, and Thompson in the early 1940's, which stressed that self-assertiveness, in contrast to aggression, is the normal emergence of the person's psychic potential.

In *Man for Himself* (1947), Fromm discussed the 'productive orientation' of personality and distinguished between it and

> . . . creativeness, particular artistic creativeness. The real artist, indeed, is the most convincing representative of productiveness. But not all artists are

> productive; a conventional painting, e.g., may exhibit nothing more than . . . technical skill. . . . But a person can experience, see, feel, and think productively without having the gift to create something visible or communicable. *Productiveness is an attitude which every human being is capable of, unless he is mentally and emotionally crippled.*[37]

That is, productiveness and creativity are not to be viewed as sublimations, as beneficial defense systems against sexual or aggressive/destructive drives, but as fulfillment or normal potential which all too often is stunted.

Both Fromm and Schachtel link creativity with fullness of experience, which is not a process limited to the artist. In *Metamorphosis* (1959), Schachtel designates this mode of relatedness the *creative encounter*, which he discusses in a critique of Kris's theory of creativity in the arts as depending on regression in the service of the strong ego. "The freshness, spontaneity, interest, and openness with which the object is approached and reacted to" are characteristic of the creative experience, in which the child, the artist, and the wise man are similar. The creative experience reflects the need for relatedness which in the artist is expressed in his "lifelong effort to grasp and render something which he has envisaged in his encounter with the world." According to Schachtel

> . . . the freedom of the approach is not due to a drive discharge but to the openness in the encounter with the object of the creative labor . . . [It is not the] product of a *repressed libidinal aggressive impulse* and of a regression to infantile modes of thought or experience, to the primary process, albeit in the service of the ego.[38]

Schachtel presented the groundwork for this view in his paper of 1950 in which he explained Rorschach's original theory linking the potential for "inner creation" with movement responses to the test figures. It is the artist's aliveness which is reflected in the kinesthetic responses.

> The life, the movement in which the perceiver animates the percept is his own life, his own movement. . . . By putting oneself inside [the percept] in imagination, by feeling from inside . . . how it moves and lives [one is involved in] what Rorschach called the capacity for inner creation.[39]

This ability to put oneself inside the percept Schachtel believes to be related to "kinesthetic empathy" and is one factor in the capacity for creative experience, which is not to be confused with the creative production for which it is a prerequisite. And Schachtel gave as the essential characteristic:

> . . . the peculiar aliveness which distinguishes the true work of art from other products of human skill and endeavor and the relation which exists between the artist-creator and his creation.

The central theme, then, in this line of thought is that creativity is linked with what one might call a distillation of normal capacities.

The difference between Kris and Schachtel comes about partly from the ambiguity resident in the Freudian term *regression,* and partly from the lack of consideration of the symbolic mode of communication as a process in its own right. Schachtel does not mention it at all and Kris refers to it only in the classical Freudian sense of its subserving such psychic functions as wishfulfillment.

Schachtel has emphasized openness of encounter, as has Maslow. Maslow, in the Symposium, *Creativity and Its Cultivation,* remarked that "psychology knows little about special talent of genius type" or what its correlation with psychologic health may be, and addressed himself to the kind of creativity that is the "universal heritage of every human being," and "covaries with psychological health." He distinguishes between the conventionally applauded creative products and the unsung creators such as the housewife with originality and resourcefulness in her family relationships and homemaking. Such creative people are described as being childlike in their openness to experience, and their spontaneity and freedom from stereotypes which parallels how the creative artist functions. Their creativity "seemed to be an epiphenomenon of their greater wholeness and integration." Maslow discusses "peak-experience," the state in which

> . . . the person becomes unified; for the time being, the splits, polarities, and dissociations within him tend to be resolved; the civil war within is neither won or lost but transcended.[40]

This concept is reminiscent of sublimation but within the gestalt this state is supernormal. The pivotal concept here is self-realization.

Rollo May and other neo-Freudian analysts have emphasized creativity as process—"basically the process of making, of bringing into being."[41] Carl Rogers uses novelty as an additional criterion for creativity:

> Creative process . . . . is the emergence in action of a novel relational product, growing out of the uniqueness of the individual on the one hand, and the materials, events, people, or circumstances of his life on the other.[42]

Rogers does not think the difference between artistic and scientific creativity particularly important, which, in effect, is ignoring the distinctive character of the symbolic mode of communication. The soundness of his criterion of "novel relational product" is questionable, since it does not differentiate what we experience as a meaningful creative product from mere oddity. Glenway Westcott, novelist and essayist, pointed out the limitations to art of extreme innovation, in the following statement:

> Originality, that is to say, individuality—that mere uniqueness which distinguishes one face from every other face, one life history from every other (likewise any given sunset and leaf and bird and bug from all [others] that are almost identical, but not quite)—is of greater importance than any extreme innovation, or one-man revolutionizing effect, or disturbance of the art for disturbance's sake.[43]

The affirmative view on creativity is best expressed by Ghiselin, poet and critic, in *Creative Process* (1952), which reports the subjective experience of people of major talent: "The creative process is the process of change, of development, of evolution, in the organization of subjective life."[44] Yet this statement, as several of the above, does not come to terms with what characterizes the artist's enterprise, nor with the recognition of the creative art-product.

A review of the formulations about creativity by the neo-Freudian analysts indicates that while creativity in general has been discussed, artistic creativity has hardly been touched on nor has psychodynamics in connection with creativity been explicitly considered. The latter may well be due to the concept of the neo-Freudians that creativity is a normal potential and that the psychodynamics suggested by Freudians derives from an erroneous view of human nature. The work remains to be done.

Other Approaches

Another of the contemporary psychological approaches to creativity is "operational" and in terms of "problem-solving." Indeed, computers have been programmed to be a model of the concept that creativity is problem-solving and have turned out musical pieces and visual abstract productions said not to be lacking in esthetic value.

In a symposium published in 1963 under the title, *Contemporary Approaches to Creativity,* several attempts were made to define creativity and to set up criteria for its recognition in order to permit the development of experiments to explore creativity. One criterion suggested by Bruner is based on the impact on the observer: "An act that produces *effective surprise*—this I shall take as the hallmark of creative enterprise."[45] Another formulation of creativity appears to depend on a reductionist sleight-of-hand reasoning which eliminates the very creativity we are interested in and, by means of the following steps, again arrives at the equation creativity=problem-solving:

First:

> If we reserve the term "creative" for activities such as the discovery of the special theorem of relativity or the composition of Beethoven's Seventh Symphony, then no example of a creative mechanism [meaning machine] exists at the present time.

Second:

> If we are willing to regard all human complex problem-solving as creative, then . . . successful programs for problem-solving mechanisms [computers] that simulate human problem-solvers already exist.

Third:

> Creative activity appears simply to be a special class of problem-solving activity characterized by novelty, unconventionality, persistence and difficulty in problem formulation.
>
> To assume that creative thinking is a kind of problem-solving . . . seems to us a useful working hypothesis.[46]

Not only does this approach explicitly omit from consideration major talent in both the arts and sciences but totally ignores the symbolic mode of communication. Granted the importance of problem-solving, and that it involves its own kind of creativity and intelligence which is ancillary to the creative process in the arts; and granted the intelligence and creativity of the men behind the computers, we are still left with the enigmas of Praxiteles, Euripedes, Michaelangelo or Beethoven.

Of the various points of importance about art and artists touched on, the most fundamental is the nuclear role of the symbolic mode of communication for both artist and audience in functioning which is not axiomatically correlated with the pathological. The contribution made by philosophy to the understanding of this mode of communication is one outstanding instance of resources available to psychoanalysis from fields other than our own.

Clinical Considerations

Though it is quite apparent that analysts of the stature of Kris write on the basis of extensive clinical experience with artists, systematic case reports are, understandably, infrequently presented. One series is Bergler's in support of his theory concerning the determinants of the artist-disposition. A psychoanalytically oriented research study of artists of demonstrated potential has recently been published from the Postgraduate Center for Mental Health, under the title *Artistic Productivity and Mental Health.*[47] The study addressed itself to the question: How would personality change brought about by therapy affect the artist's work? The distinction was made between the phenomenon of creative talent and its mobilization into productivity. The checkpoints were based on behavioral patterns, that is, on work habits, rather than on the quality of the art-products, which would have introduced more variable personal judgment. The conclusions drawn were that in the six artists cited there existed certain connections between specific personality patterns and flaws in the work pattern, and that psychotherapy, in addition to improving their personality and sense of well-being, helped them to more effective work-habits. That is, "the artist who gets emotionally well [does not] cease to create"; rather, he creates more effectively.

The distinction made between creativity and productivity is comparable to that between biologic conception and bringing the pregnancy to term. But the creative process as a whole certainly involves the delivery at term of the art-product. That is, productivity is one stage in the creative process.

We can take as axiomatic that there are many whose creative potential does not become activated. To realize one's capacity for artistic creativity requires selection among available processes within oneself. As the heroine in *Marathon 33* put it, *you must respect it if you've got it; you must love and cherish it.* One group in my clinical experience consists of those of validated talent who renounced their allegiance to it and turned to the service professions—a pianist who became a physician, a singer with some talent for composing who turned to social work, a gifted writer who became a teacher of clinical psychology. These three felt that their pursuit of their creative potential in art would be selfish self-indulgence without utility to the needy. Their redirection of professional purpose was based on a sense of guilt about "self-indulgence" and a lack of conviction in the value of intangibles.

The committed artist experiences his sense of self and identity in the creative process. One painter expressed this as follows: "When I am not painting, I feel I am not doing anything at all. I have wished at times I could give all my energy to painting, but then I would not be living. The Japanese have an interesting attitude—the artist is almost a priest."

The vividness of the experience of imagery constitutes the artist's *creative vision*, both for the visual and the literary arts. As the painter described it, "When I am feeling good I can get to put my images down on canvas. I always have images—one now is of a seated woman in brilliant reds, blues. From being on the beach in the summer, I still have images now—white, glitter, umbrellas and the sea." For one writer the fiction characters became full of such visual vividness that he could see the changes in their expression.

Not only is the artist dedicated to achieving the art-product, but he has a sense of wanting to say something by means of it. If that is missing, either he stops producing or becomes a stereotype of himself. Perhaps E. M. Forster illustrates this kind of loss of the drive to communicate. When asked what accounted for his not writing any more novels in the last forty years, he said he abhorred the Machine Age and had written to protest its coming, but now that it's a fact there's not much point to further written protest.

Certain kinds of despair and depression which remove the sense of goal and of meaning in living are particularly destructive to creativity; retardation in depression may attenuate the progression to the goal. When depression constricts image- and symbol-making as well as the freedom of verbal communication, the very stuff that dreams and works of art are made of is missing.

It requires stamina to tolerate the changing phases of the creative state with its fluctuant stresses. The subjective experience in both the precursor tension and the active state of productivity is shunned by some who prefer not to mobilize their creativity at this cost. Two patients of artistic promise, who had had schizophrenic episodes, shunned the cultivation of imagery and systematic artistic pursuit out of a painful association of fantasy processes with the schizophrenic experience, and a distrust of the ability to avoid one.

The artist's relationship to time, his style of using time, affect the working out of his creativity. Some need the sense of pressure of working against a deadline; one writer of fiction needed the sense of unbounded time and of freedom from the psyches of others which he got by working in the depths of night. Cutting off work before he is ready disrupts the artist's productivity and the train of thought and feeling, leaving a sense of frustration, if not actually a loss of production,—the syndrome of the Man from Porlock.

An element close to the core of the creative process is the artist's *use of life experience.* This is most easily recognizable in creative writers. To cite a few instances. One writer worked through nightmarish experiences of violence in adolescence into fictional form, arriving at meanings and completing the experience of pain and resolution, and not with wishfulfillment in the confines of the experience. Actually a crisis of identity was involved and resolution of illness was achieved through creativity. A second wrote a story in which the hero "walked out" on his wife, which was the writer's own wish which he had decided against—reminiscent of Freud's artist who makes the excursion into fantasy and back to reality, without taking the fantasied action. O'Neill is said to have written *Long Day's Journey into Night* out of an inexorable need to work through and make sense of the tragic family experience. Parenthetically, this working through of life-experience as one element in the creative process is not limited to artists. We analysts often use our own life-experience as an intuitional resource for patients. One writer, who attained a mood of affirmation on emerging from a prolonged depression, conceived of fictive characters who were not closely identified with himself or people in his life, but with a great mythic theme of disaster and affirmation, a transposition into the world of fiction of his own trust of affirmation.

These various experiences with artists indicate that their creativity involves bringing their experiences in living into focus in the art-product. The work of art enhances our experience and vision while it reflects the artist's.

Thesis: Though the symbolic mode of communication may be involved in illness, it is essentially a normal and vital human function, heightened in both the creativity in the arts and their appreciation. Art will always be happening just as we have joy and conflict, rage and doubt and grief and transcendence, whether experienced in the course of living or through the communication of art.

REFERENCES

1. Greene, B.: The artist's reluctance to communicate. Paper read in Symposium, The Artist and His Milieu, at New York, Postgraduate Center for Mental Health, Dec. 12, 1956.
2. Dewey, J.: Art as Experience. New York, Capricorn Books, G. P. Putnam's Sons, 1958.

3. Barzun, J.: The Energies of Art. New York, Vintage Books, 1962, p. 323.
3a. Brain, L.: Population plan for world urged. New York Times, Aug. 27, 1964, p. 35.
4. James, H.: Preface to The Aspern Papers. *In* The Art of the Novel, R. P. Blackmuir, New York, Charles Scribner's Sons, 1934.
5. Shaw, G. B.: Postscript to Back to Methusaleh, cited by M. Henle, from The Birth and Death of Ideas, H. E. Gruber, G. Terrell, M. Wertheimer, eds. *In* Contemporary Approaches to Creative Thinking, New York, Atherton Press, 1963, p. 40.
5a. Ciardi, J.: Manner of Speaking. Saturday Review 46:23, 1964.
6. Tolstoy, L.: Quoted by H. Read, The Meaning of Art. Baltimore, Penguin Books, 1963, p. 191.
6a. Dewey, J.: *Op. cit.*, p. 80.
6b. Spiegel, R.: The Experience of Art. J. Am. A. Univ. Women 53:231, 1960.
6c. Aristotle: Butcher, S. H., ed. Aristotle's Theory of Poetry and Fine Arts. New York, Dover Publications, 1951, p. 240.
7. Nordau, M.: Degeneration. New York, D. Appleton, 1895. Translated from the German.
8. Breuer, J., and Freud, S.: Studies in hysteria. *In* Nervous and Mental Disease Monographs, New York, Coolidge Foundation, 1947.
9. Freud, S.: The interpretation of dreams. *In* The Basic Writings of Sigmund Freud, A. A. Brill, ed. New York, The Modern Library, Random House, 1938, pp. 309 ff.
10. ——: Delusions and dreams. *In* Jensen's Gradiva, P. Rieff, ed. Boston, The Beacon Press, 1956, p. 33.
11. ——: The poet and daydreaming. *In* Collected Papers, vol. 4. London, Hogarth Press, 1946, p. 183.
12. ——: Sexual morality and nervousness. *Ibid,* vol. 2, p. 82.
13. ——: Civilization and its discontents. The Standard Edition, vol. 21. London, Hogarth Press, 1961, p. 79.
14. ——: Mental Functioning. *In* Collected Papers, vol. 4. London, Hogarth Press, p. 19.
15. Malraux, A.: Voices of Silence. New York, Doubleday & Co., 1956, p. 419.
16. Freud, S.: Dostoevsky and parricide. *In* Collected Papers, vol. 5. London, Hogarth Press, 1950, p. 222.
17. ——: Contributions to the psychology of love. A special type of choice of object made by men. *Ibid.* vol. 4, p. 192.
18. ——: Totem and taboo. The Basic Writings of Sigmund Freud. A. A. Brill, ed. New York, The Modern Library, Random House, p. 877.
19. ——: The theme of the three caskets. *In* Collected Papers, vol. 4, London, Hogarth Press, p. 255.
20. ——: Character types in psycho-analytic work. *Ibid,* p. 339.
21. ——: The moses of Michelangelo. *Ibid,* p. 258.
22. ——: Our attitudes towards death. *Ibid,* p. 307.
23. Brill, A. A.: quoted by Bergler, E. The Writer and Psychoanalysis. New York, Robert Brunner, 1954, p. 12.
24. Klein, M.: The Psychoanalysis of Children. New York, Grove Press, 1960.

25. Bergler, E.: *Op. cit.*, p. xii.
26. Bychowski, G.: The metapsychology of artistic creation. Psychoanalyt. Quart. 20:592, 1951.
27. Kris, E.: Psychoanalytic Explorations in Art. New York, International Universities Press, 1952, p. 20.
28. Kubie, L. S.: Neurotic Distortion of the Creative Process. Lawrence, University of Kansas Press, 1958.
29. Rank, O.: Art and artist. *In* The Myth of the Birth of the Hero and Other Writings, P. Freund, New York, Vintage Books, 1959, p. 211.
30. *Ibid*, p. 243.
31. Arieti, S.: The Rise of Creativity: From Primary to Tertiary Process. Contemp. Psychoanalyt. 1:59, 1964.
32. Freud, S.: See reference 9, p. 509, 525ff.
33. Bleuler, E.: Dementia Praecox or the Group of Schizophrenias. New York, International University Press, 1950, pp. 67, 374.
34. Fenichel, O.: The Psychoanalytic Theory of Neurosis. New York, W. W. Norton, 1945, p. 48.
34a. Kris, E.: *Op. cit.*, p. 48.
35. Fromm, E.: The Forgotten Language. New York, Rinehart & Co., 1951.
36. Tauber, E. S., and Green, M. R.: Prelogical Experience. New York, Basic Books, 1959.
36a. Spiegel, R.: Specific Problems of Communication in Psychiatric Conditions. *In* American Handbook of Psychiatric, S. Arieti, ed. New York, Basic Books, 1958, p. 909.
36b. Thompson, C.: Psychoanalysis: Its Evolution. New York, Hermitage House, 1959, p. 59.
37. Fromm, E.: Man for Himself. New York, Rinehart & Co., 1947, p. 85.
38. Schachtel, E. G.: Metamorphosis. New York, Basic Books, 1959, p. 241ff.
39. ——:Projection and its relation to character attitudes and creativity in the kinesthetic responses. Psychiatry 13:92-93, 1950.
40. Maslow, A. H.: Creativity in self-actualizing people. *In* Creativity and Its Cultivation, H. H. Anderson, ed. New York, Harper & Row, 1959, p. 89.
41. May, R.: The Nature of Creativity. *Ibid*, p. 58.
42. Rogers, C. R.: Toward a Theory of Creativity. *Ibid*, p. 71.
43. Wescott, G.: Images of Truth. New York, Harper & Row, 1962, p. 6.
44. Ghiselin, B.: The Creative Process. New York, Mentor Book, The New American Library of World Literature, 1952, p. 12.
45. Bruner, J. S.: The conditions of creativity. *In* Contemporary Approaches to Creativity. New York, Atherton Press, 1963, p. 3.
46. Newell, A., Shaw, J. C., and Simon, H. A.: The processes of creative thinking. *Ibid*, p. 65, 66; order of quotations varies from text.
47. Fried, E., *et al.*: Artistic Productivity and Mental Health. Springfield, Ill., Charles C Thomas, 1964, p. 163.

THE WAR WITH WORDS: CREATIVITY AND SUCCESS*

By HARLEY C. SHANDS, M.D.

God made a little gentian;
It tried to be a rose
And failed, and all the summer laughed.
But just before the snows
There came a purple creature
That ravished all the hill;
And summer hid her forehead,
And mockery was still.
The frosts were her condition;
The Tyrian would not come
Until the North evoked it
"Creator! Shall I bloom?"

Emily Dickinson

WHEN HUMAN BEINGS first become interested in origins, there is a tendency for explanation to use the model of naive experience: Athena, the goddess of wisdom is said to emerge fully-armed from the head of Zeus with as little preparation as there apparently is when insight dawns in the experience of the "*Aha-Erlebnis*." With the suddenness of creation there goes the notion of a powerful agent, powerful in the way human beings wish that they themselves were powerful: God says, "let there be light," and there *is* light. Even in the nineteenth century, remnants of these attitudes persisted in the resistance to the theory of evolution and in the durable nature of the idea of spontaneous generation.

Systematic investigation has by now left little doubt that creation

---

*Supported in part by NIMH Grant MH 05518-03 and the Commonwealth Fund.

The quotation from Emily Dickinson used in this article is from *The Poems of Emily Dickinson*, edited by Martha Dickinson Bianchi and Alfred Leete Hampson, and appears through the kind cooperation of the publisher, Little, Brown & Co., Boston, 1939, to whom thanks are extended.

is evolution, and consensus has firmly established that natural selection is the principal mechanism by means of which mutational change is perpetuated and spread throughout the species. An extension of the selection theory has recently been influential in genetics, in the idea of "clonal selection." I would like to suggest here that the apparent miracle of artistic creation may similarly be susceptible to sober mechanistic explanation. The development of artistic "creations" appears to be based upon a proliferation of mutation with a subsequent rigorous process of selection. Mutational novelty here as in evolution may well be the effect of chance, and creativity becomes largely a matter of the selection of some, with the rejection of the vast majority, of the mutational forms. Of particular interest to me in this context is the formal similarity between this social process and the professional process of free-association and interpretation in psychotherapy.

When we speak of "insight" in either the artistic or the psychiatric context, we tend to think of a sudden enlightenment, an abrupt illumination of a previously "dark" area. We characteristically use visual metaphors, and we often think of the process as a dramatic one, e.g., the revealing of a spectacle when the curtain is raised in the theater. However, everyone involved in the preparation knows that raising the curtain is the culmination of a very long period of preparation; the more efficacious the preparation, the less apparent is it in the final spectacle. The work of genius, as Edison prosaically put it, is often 2 per cent inspiration and 98 per cent perspiration.

Anyone involved in a creative enterprise is familiar with the sequence from the first exciting gleam of an idea through a long tedious effort to shape some medium into a form which will convey the meaning so excitingly and "immediately" presented in the first glimpse. But he also knows that the original idea will suffer many important revisions as it encounters the "resistance" of the medium. The artist often feels, and may say, that the medium seems to be possessed of devils; it "fights back" with malignant and heartbreaking apparent "autonomy." So intense is the "war with words" in the case of the writer or the fight with paint and brush in that of the artist that the creative person may describe the preliminary period of setting up the typewriter, cleaning brushes, stretching canvas, and the like as positively delightful because it delays the eventual struggle with the resistance of the medium and the blankness of the surface prepared to receive the form as created.

In the extended and effortful progress toward the final construction of an expressive shape, the artist participates in a creative process which resembles that of the scientist searching for discoveries. As Kuhn points out in his book on scientific revolutions, we think in science as well as in other contexts of a miraculous revealing of discovery to the single discoverer, only to find that when we look at the historical facts, the discovery is always an event extended in time. He cites particularly the case of the discovery of oxygen, noting that the apparently momentary discovery actually was extended to several decades, occupied the energies of at least four principal investigators and finally was a summation of events taking place in several different countries. The final "discovery" in science is more like a cooperative collage than it is like a two-second Polaroid snapshot, and its ultimate consensual shape depends as much upon the "audience" composed of the consensus of scientists as it does upon the ideas advanced (to be selected or rejected) by the original investigator.

The artist or the writer similarly has to go through an extended period of creativity or discovery before his work is recognized. The creativity is often only retrospectively apparent. In the case of Miss Dickinson, whose poignant poem is quoted above, her contribution to English literature remained almost completely hidden during her lifetime. Posthumously, it was published to very modest acceptance, and only after many decades did it finally find the kind of "selection" which now establishes her as a major poet.

The work of the artist is but the other side of the necessary work of the audience; there is required as much education and effort in the latter as in the former context. The ultimate "creation" or "discovery" is necessarily a consensual matter. The goal in both art and science is a goal of "meaning," but meaning is always a matter of mutual participation. To use what may be a neologism, the artifact or the scientific idea is always a "communicandum" defining artist and audience only with reference to itself, or, to put it in other terms, a "transaction" only with reference to which does the designation of a "first party" and "second party" apply. The curious process is one in which the creator evokes not only his creation but his audience, and the audience, participating throughout, shapes the emerging work through its acceptance and rejection, in part and in whole. It is the very intermeshing of the one with the other aspect of the process which makes it so difficult to deal with in the discursive methods traditionally used in science: it is impossible to find "controls" because every change in the one implies a change in the other and the two occur simultaneously and inextricably intermingled.

The creative transaction, involving two participants, takes place in a setting of human relatedness; the background of humanness is inevitably to be taken into account in dealing with the idea of significance in symbolic terms. But for the moment to the artist it is necessary that this background of human relatedness disappear into unconsciousness as he concentrates all his attention upon the "battle" in which he is engaged. The "repressed" returns, however, in the consideration that the measure of the artist's eventual success is his ability to use himself as the principal representative of the eventual audience. By the time the successful artist has come to satisfy himself, he has created an artifact which will eventually satisfy a wide audience. The creative artist in his intense preoccupation with himself is in communion with the most significant representative of the wider human group, even if perhaps only the wider group of a later epoch. In the successful artist, the narcissistic conviction of his own superiority comes to be positively reinforced by the consensus.

The artist must be able to focus a fiercely concentrated attention upon the means of his task with only secondary attention given to the ultimate goal. He must, to put it in a form familiar from another context, "keep his eye on the ball." To the creative artist the compulsive struggle with the means of his work often occupies the stage to the total ignoring of the eventual goal. The means becomes intensely conscious to him whereas the ultimate goal may be for long periods of time quite ignored.

It is in this context that one sees the necessary counterpoint of artist and critic most clearly. Whereas the artist needs to be principally concerned with the resistance of the medium to his own effort of self-expression, the critic has constantly to see the work of the artist in the perspective of the artist's other work and in that of the relation of this artist to his contemporaries and predecessors. The artist must first and foremost respond to his own critical appraisal of himself, whereas the task of the critic is to predict the degree of general acceptance or "selection" of the artist's work in the present and in the future.

There is an interesting lag always observable between the production of an artifact or series of artifacts and their critical acceptance. At the beginning of the process, the artist is likely to find acceptance only very close to home: the "Fauves" were almost alone in their admiration of each others' paintings, and it was in the face of widespread critical hostility that their show was presented to the public. On the other hand, to the creative artist, the point of acceptance by the audience may be that at which the artist himself has already lost interest. The painter Duchamp, whose "Nude Descending the Stairs" was the sensation of the Armory show in 1913 recently said, using his own painting as an example, that a painting tends to lose the "aura" of contemporaneousness quickly. As it loses its "smell of the moment" it becomes an item of interest to the historian rather than to the critic. The poet Eliot says in a similar vein that "one has only learnt to get the better of words/ For the thing one no longer has to say, or the way in which/ One is no longer disposed to say it."

The time elapsing between the illumination of the creative leader and the beginning of his spreading of the message to a group of followers is called the period of "withdrawal-and-return" by the historian Toynbee. Leaders of this type (including Buddha and Mohammed) retire from the world for a period in which their ideas mature into a form in which it becomes possible to communicate them. During this period the effort which such a man makes is that of stabilizing a dialogue between two points of view. The one is that of the dissident, the other that of the spokesman for the culture. The creative resolution of the tension in such a system is that of learning how to express the dissident message (with modifications) in a form such that it can be communicated to the reference group to which the message is ultimately directed. This process appears to be similar to that in which the scientist retires to his laboratory to work out the implications of a hypothesis, or the artist to his studio to accumulate enough work to interest a gallery in giving him a show. But the specific point to which I want to call attention is the degree to which such a process of withdrawal and return resembles the psychotherapeutic enterprise, perhaps most prominently in Freud's own self-analysis.

It is pertinent to the terms of the present discussion that we find in Freud the intense concentration upon a narrow field characteristic of the creative person, and that in him there occurred over a very long period of time an extended process of repeated change in his "discovery." We can trace "free association" back to Galton's description decades before Freud's, and the continuity of discovery continues in Freud long after the original statements. The selection process is obvious as well, since in the institutionalization of the discoveries, there has been a prominent rejection, especially in the American scene, of Freud's repeatedly stated preference for a nonmedical rendering of the psychotherapeutic task, and as well a widespread rejection of his idea of the "death-instinct" and a more covert rejection of the late reformulation of the "hydraulic" theory of anxiety. Similarly obvious rejections of parts of the message of the master can be found in any institution based upon a dogma: witness the deviations of the Christian Church and the Communist Party from their respective origins.

The general view of the creative process acting by mutation and selection in a context in which "many are called but few are chosen" serves to point up the specificity of application of the same idea to the more mundane process of psychotherapy. I want to suggest here that "free association" is primarily a technique oriented toward the emergence of novel comments within a social situation; it may not be necessary to note that these comments are often novel only in the patient's own experience of himself—the therapist comes to feel sometimes with the ancient Solomon that there is nothing new under the sun, but the patient is characteristically astonished. In a biological analogy, free association may have some resemblance to the use of a mutation-facilitating agent such as colchicine in plant-breeding. From the flow of material fostered in free association the therapist carries out a continuous process of selection by positive and negative reinforcement. The process is one of "therapeutic selection" analogous to the process of social selection in which all human groups facilitate the child's use of certain phonemes and sequences of phonemes while suppressing vastly greater potential numbers of phonemes and sequences.

It becomes less difficult to understand the protracted nature of the psychotherapeutic enterprise when we learn comprehensively to understand the tedious and extended nature of the process of discovery. If it took several decades and the work of many different laboratories to "discover" oxygen, a relatively simple element, it makes more sense that it should take months or years to discover the self. The process is one of an extended dialectic the predictable outcome of which is that the "self" discovered will tend to resemble the institution described by the therapist. The process is much more one of mutual participation in the creation of an artifact than it is one of "cure" in any simple medical analogy. The opinions of the treated, as Glover emphasized many years ago, tend to resemble in a remarkable degree the opinions of the treaters; we can say that the opinions of the fathers are visited upon the sons even to the third and fourth generations.

The process of selection emphasizes the reciprocity of explanation. To the human being looking back through evolution, the progress upward toward the emergence of his own form gives every evidence of being directed, creative, and praiseworthy, but on the other hand, the now quite conceivable possibility that human beings might in some future time be superseded by descendant self-reproducing machines appears horrid in the extreme. From a detached point of view, however, one could not argue that any species has a corner on perpetuity, and it is certainly true that chance is likely to govern future evolution as we painfully come to recognize that it has governed it in the past.

In the creation of a species, of a faith in a prospective convert, of a self-concept in a patient, or of a theory or artifact in the intellectual investigator or artist, we tend to the same kind of reciprocity of understanding. The convert struggles against submission, but once having surrendered, he proclaims that he has prevailed! And so paradoxical are the processes of socialization that he is to a considerable extent right. The submission of Saul was the crucial event in the conversion of Saul into Paul and the creation, in Paul's image largely, of the Christian Church as eventually institutionalized.

In the technique of psychotherapy, both human relatedness and focusing are significantly involved. First, with relation to focusing upon the particular immediate resistance to expression, the therapist repetitively insists that the patient pay attention not to long-range goals but to focusing upon the next few words which "come to him." These words present embryonic ideas and the therapist in his role as a critic (and it must be remembered that in the modern world of art a friendly critic is the artist's best ally) insists upon a continuous stream of such partial artifacts which he subtly (and continuously) influences by his responses. The therapist's technique is a combination of methods for facilitating the flow and modifying its form. The therapist naturally emphasizes those particulars which he can most easily fit into his own understanding, just as the critic tends to approve of those artifacts which fit into his own theory of art. A major problem in both art and psychotherapy is that of allaying the anxiety of the producer so that he can bear to expose enough of his work to the potential critical view of the respondent for this latter to get an impression of a whole.

A major technical approach in psychotherapy is very similar to that of the critic, namely, that the psychotherapist insists that the patient demonstrate the particulars of his experience; he asks "when?" "where?" "with whom?" and "exactly what happened?" In this way he insists upon making a judgment in terms of the series of "artifacts" which can be shown to him by the artist. It is especially in "analyzing the transference" that we see the significance of this kind of demand, with the further implication that every such particular comment made now with reference to the therapist has a general applicability to categories of persons in the patient's experience. It is only when the therapist can make the generalization on the basis of the patient's particular comment that further understanding emerges. Much of the most powerful resistance of the patient is his reluctance to present such particulars, and on the other hand, his unwillingness to believe that his specific relation to the therapist is a good basis upon which to generalize to other human relationships. Again it may be noted that the arrogation to oneself of the evaluative function is one of the principal modes of claiming superiority or dominance in human relationships. Only by submitting to this demand of the critic-therapist can the patient-artist both submit and prevail, but it is perhaps worth mentioning that critics and perhaps even therapists as well have been known to be wrong!

The central mutuality of the psychotherapeutic relationship is in participation toward the construction of a durable self-concept, one in which contradiction is either diminished or specifically tolerated in some enduring state. If one understands a self-concept not as based upon historically accurate truth (which is in point of fact quite impossible because of the relentless erosion of "truth" over the years) but as an artifact, then it is easier to understand some aspects of the techniques as they are known to competent psychotherapists. In one of these contexts, for instance, we find psychotherapists in general negatively reinforcing the making of notes; in a similar instruction, Proust says that if a prospective writer collects notes he is lost. On the other hand, along with experienced psychotherapists, Proust points out that every detail of every character appearing in any novel has to be a detail derived from the author's acquaintance with someone. The important consideration is that the detail as appearing in the novel or in the self-concept appears as a bit of a spontaneously emerging mosaic, greatly modified by its relation to other neighboring details and the whole.

The specific anxiety of the creative artist is very effectively reproduced in the psychotherapeutic situation; both the patient in free association and the artist in his studio face a terrible void, a blankness which looms with horror before him. The psychotherapeutic situation utilizes both the *horror vacui* and its specific "medication," a human relationship, as the therapist partially fills in the void by his support and encouragement while at the same time (and unconsciously) forcing the patient to move forward into anxiety. I say unconsciously here because many a therapist appears not to recognize the immense coercion implicit in his definition as spokesman for the relevant human group. The limiting factor in the selection of patients for intensive free associational psychotherapy appears to be inherent in this process: it is only the patient who is able to contemplate months of blankness or void as he creates himself with a minimum of outside help that is a good prospect for therapy.

The help which is essential comes, of course, from the patient's own previous history. With good luck, the suitable patient has been able so to internalize a world that he can endlessly bring up tentative forms with which to fill the yawning chasm which confronts him as the therapist says, "Just say whatever comes to mind." The patient responding successfully to this instruction to some extent resembles Fortunatus whose purse was never empty, or the widow whose cruse flowed oil endlessly upon demand. It is only when this basic requirement can be at least minimally met that the therapist can use his selective techniques with the greatest effectiveness.

When the patient has the capacity and the courage to proceed in this creative enterprise, he regularly finds to his own astonishment that the apparent void, like that of the blank canvas, is filled with potentialities of structuring, both in terms of the plasticity of the medium and the acceptance of experiment on the part of the relevant human representative. To the creative artist the client and the critic present a resistance which resembles that presented to the patient by the therapist. If the artist, hopeful of pleasing, says to the client, "What shall I make?" the sophisticated client often dismisses any possibility that the artist might be great, since the great artist by definition knows what to construct. On the other hand, when the artist goes ahead to construct something which expresses his own feeling, more often than not the client then responds with, "Not that!" Many of the most eventually successful of artists have for years painted pictures which met with only the reinforcement of a powerful self-approval, eventually (but sometimes only after the artist has long since died in a madhouse as Van Gogh did) to find a consensual selection and acclaim.

The similarity of this process to the psychotherapeutic one repeatedly presents itself when one looks for it. The therapist's silence has an effect like that of the unwillingness of the client to buy or of the critic to approve. To the patient struggling in isolation to bridge the gap separating him from his principal critic, the silence of the therapist says, "Not that!"

Looked at in this way, the notion that the psychotherapist is occupied in working with the patient toward the creative construction of the primary artifact of human beings, a self-concept, appears more appropriate than the often preferred notion that the psychotherapist is concerned with "curing" a disease. The process bears a much greater resemblance to the selective reinforcement of certain modes of action with the selective discouragement of others characteristic of the activity of an athletic coach or an instructor in art school than it does to the surgical approach to an unconscious body. Reference to the similarity in the creative approach in both instances tends also to suggest a possibility rarely or never addressed in psychotherapeutic circles, namely the possibility that sometimes the artist is more perceptive than the critic. Perhaps more often than we would like to believe the patient may be more creatively than pathologically deviant, and we may in fact be working with a process which so powerfully stamps out deviance that we may be selecting only the commonplace. But perhaps we can console ourselves in this regard that if the deviance is that of a truly divine discontent it will be proof against therapy!

The essence of a creative process appears repeatedly to be that of filling in the void which separates human beings from each other. Paradoxically, as the void is filled in, there occurs, along with the relating of the one to the other also a separation of the one from the other. It is as though the structuring of the various "media" through which human beings know each other is the essential preliminary to the separation which necessarily precedes choosing. The severely disturbed patient cannot communicate because he cannot tolerate nothingness long enough to demonstrate to himself the repetitive miracle of human socialization, namely that in the void there is always the potentiality of structuring and that human beings are inevitably disposed, insofar as they are human, to the consensual formation of a mutual structure from which each emerges to the other in the repetitive miracle of coming to know each other.

Finally, it seems important to emphasize the thesis that creativity is not unusual, it is in fact the most routine of facts in biology. Evidence from a variety of sources suggests that the human body is constantly being created anew from day to day, in accordance with "instructions" coded into the proteins of the protoplasm. Similarly, in ideas such as that of the reconstruction theory of memory proposed by Bartlett thirty years ago and recently supported by new evidence, it would appear likely that thoughts and ideas are extravagantly recreated in a stream endlessly flowing from birth to death. The basic fact is not that of creativity but that of selectivity when we seek to understand the emergence of "new" ideas and novel artifacts.

We find in successful therapy and in successful art that after a long period of effort each participant appears highly pleased with the activity of the other, and the void which was once so frightening now appears to be the site of a communicational structure which fills it in and leaves no "hole" at all. The gap is filled with a structure of meaning, a structure which curiously enough has no concrete "existence" since it is purely formal in the most abstract sense. The shared structure is, to turn to computer language, a shared program, and its significance is in many ways a purely ritual one, in that its meaning is simply a vehicle for participation between the members of the system.

The essence of the creative process is that of the filling in of the void which separates human beings from each other; the means of filling this void in the human group is that of the formation of a structure out of nothing, but the means by which the structure is accomplished is that of the consensual selection, out of all possible forms, of those which can be the focus of substantial agreement. The original, the novel, or the "creative" person is differentiated from his fellows primarily by his ability to anticipate future structurings; to some extent his work is a self-fulfilling prophecy in that it strongly colors the eventual acceptance, but in another way it is entirely at the mercy of the consensus. The incredible trick which the creative artist is able to accomplish is that of the successful struggle with his medium in such a way as to anticipate the eventual acceptance of it by the group. He paints the future! but the future so painted will not come into its own until it is selected by the group. In a precisely similar way, the psychotherapist acting as an audience encourages the patient as a potentially creative person "working in life," to throw a bridge across the void separating them, but so persistent and so selective is the activity of the psychotherapist that he tends to constrain the eventual form of this bridge so as to conform more or less to that in which he has previously himself been trained. In human life, as the French say, "*Plus ça change, plus c'est la même chose!*"

REFERENCES

BARTLETT, F. C.: Remembering. Cambridge, London, 1932.

BURNET, F. M.: Immunological recognition of self. Science 133:307, 1961.

DUCHAMPS, MARCEL: Interview by William Seitz. Vogue Feb. 15, 1963.

ELIOT, T. S.: Four Quartets. N.Y., Harcourt Brace, 1943.

FESTINGER, L.: A Theory of Cognitive Dissonance. Stanford, Stanford University Press, 1962.

GALTON, FRANCIS: Psychometric Experiments. Brain 2:149, 1879.

KUHN, T. S.: The Structure of Scientific Revolutions. Chicago, Chicago University Press, 1962.

PIAGET, J.: The Origins of Intelligence in Childhood. N. Y., Norton, 1952.

SHANDS, H. C.: Complementarity in psychodynamic theory. *In Science* and Psychoanalysis, Vol. V, J. Masserman, ed. New York and London, Grune & Stratton, 1962, p. 66.

Part II. The Community

Psychoanalysis and the Community

INTRODUCTION

By JOEL ELKES, M.D.

SOME FORTY YEARS AGO there began on this continent a dialogue between psychoanalysis and various sister sciences which, on the basis of long and hard-won tradition, claimed a knowledge of man, his origins, social habits and institutions. In its short life, the Academy of Psychoanalysis has done its full share in furthering and amplifying the liaison. Indeed, it has become a conversation not only between psychoanalytic theory and psychotherapeutic practice, between the uses of the dyadic transaction and the formative powers of a social field, but fundamentally between a biological determinism rooted in the immutability of drives and their destructive dilemma, and a social determinism in which social learning can further individual growth and social responsibility. To pit individuality against the web of our present-day togetherness is quite a trick, yet it is not beyond the verve which so characteristically and significantly propels our American society toward a more charitable, kinder and deeper image of man.

There is, of course, danger in this very optimism. The principal one is superficiality and its first cousin irresponsibility. Patness in explanation; doctrine half understood, yet trebly enforced; second handedness, and relegation to the past—these are some of the dangers. But we can take our modest place in Pediatrics, in Medicine, in Geriatrics, and what we already know is relevant to neighborhood programs, to schools, to churches, to courts, to police and to the planning of new communities. We assume our role in society by making a deliberate effort to return to society what belongs to society; we can earnestly and steadily reduce the need for our own services by placing with the family what is the family's, with the school what is the school's, with the churches and the other social institutions what belongs to them. As members and agents of society in mental health centers, we could enhance their trust in us; and trust, like peace is indivisible.

THE PSYCHOANALYST AND COMMUNITY PSYCHIATRY

By STANLEY F. YOLLES, M.D.

IT IS MY CONTENTION, in continuing this section, that the psychoanalyst can have his couch and his community too. And, more importantly, that the United States cannot have a truly comprehensive community mental health program unless the psychiatrist whose knowledge is solidly grounded in traditional clinical psychiatry provides program leadership based on that knowledge.

All of us are well aware of the changes taking place in the treatment of mental illness. The steadily increasing public demand for more extended psychiatric services available to a wider segment of the population confronts us with an inescapable challenge. The psychoanalyst of today—born of yesterday's revolutionary concepts in the search for knowledge about the mind of man—cannot ignore the opportunity to use current knowledge as comprehensively as possible. And I submit that as he accepts this, he will find that participation in widespread community mental health programs becomes for him not only exciting but compelling.

There may be those who welcome the advent of community psychiatry as a blow to the pre-eminence of psychoanalysis in United States psychiatry. I cannot agree with this point of view. The psychoanalyst in maintaining the traditional one-to-one relationship with an individual patient has always been, at least indirectly, concerned with the environment in which that patient lives. The basic concepts of psychoanalytic theory, then, can and should be extended to group and community relationships.

If this is true, it is equally unsound for psychoanalysts to reject the champions of community psychiatry in the belief that the two viewpoints are incompatible. A more realistic view would be the attitude that the psychoanalyst can and should be deeply involved in structuring the new programs. For example, the psychoanalyst can help to maintain a necessary sense of proportion in the development of a community mental health program by assuring that treatment is neither too rapid nor too superficial. I do not suggest that an appropriate balance between present theories and the development of new ones will be easily achieved. But that it can and will be achieved I do not doubt.

I see no essential incompatibility between the theories of community psychiatry as they adapt to a population and the theories of psychoanalysis as oriented to the individual. The traditional clinicians and the community specialists cannot allow themselves to be divided into two groups, since the major difference between them relates only to the primary object of their professional interest.

The psychoanalyst's primary concern is the welfare of his individual patients. The community psychiatrist's concern includes not only the individual patient he sees but also all the other disturbed, or potentially disturbed, people in the population he serves. Obviously, he cannot treat them all, but he believes that he can exert a positive influence through development of additional methods of intervention. That intervention, from the point of view of community psychiatry, includes action by all professionals of the helping disciplines, as well as by key citizens in the community, to improve the physical, psychological and social environment within which they live.

It must concern all psychiatrists that we live in a society in which so many persons are emotionally disabled. In the present day all of us must have regard for the environmental factors relevant to promoting mental health, in order to reduce the pathogenicity of the community. In helping to maintain sound mental health, the psychiatrist, of course, will become involved in the social system. He can best do so by serving as a consultant to others in the community whose decisions affect that system.

The psychoanalyst can do so effectively since he brings to the problem a full appreciation of the complexities that exist between the individual and the social context in which he functions. Further, the analyst has special orientation to concepts of the levels of ego growth and socialization.

Even though many psychoanalysts accept these responsibilities, to some it seems impossible that a psychiatrist, focusing on the community and using community-oriented techniques, may affect the lives of a large number of disordered persons and still maintain his accustomed personal and professional life. However, many who have already made this adaptation in their professional practice have proved that it can be done.

I might note that most psychoanalysts who have chosen to concern themselves with community psychiatry do so on a part time basis, while retaining a portion of their private practice. From a personal standpoint, this is economically feasible for them. But in so doing, they are also able to contribute most effectively to the sound mental health treatment program.

This program will develop through collaboration among psychiatrists, psychologists and social workers on the one hand, and social and political scientists, clergymen, administrators, economists, city planners, lawyers, politicians and citizen leaders on the other. Each of these persons will participate in a mental health program for a variety of reasons, based on a diversity of primary personal concerns. The involvement of the clinical psychiatrist is based on his primary concern for the individual patient. He must, therefore, maintain his personal identification with psychiatry, and his competence in this medical specialty. Only thus can he make sure that a community treatment program is firmly committed to the welfare of each patient, even though the number of patients is greatly increased.

The people of this country, through the Joint Commission on Mental Illness and Health, the policies of two United States Presidents, and the Congress, have stated that they wish to support a network of adequate and comprehensive services for the mentally ill. It remains for the clinical psychiatrist to define his role in providing such a service constellation. And I suggest that the traditional psychoanalytical approach in which a relatively few cases are selected for treatment can and should be adapted to meet the wider need.

The psychoanalyst has been trained in observation, analysis and definition. He can use these special skills to the public advantage in consultation and community organization, since the skills are derived from understanding and competence in identifying and managing unconscious processes in himself and others.

Therefore, as communities progress in providing treatment programs, the psychoanalyst can provide a unique kind of leadership in clarifying the integration of clinical and community orientations. In doing so, he can add depth of perception to the breadth of need, as the entire medical profession moves toward the collaboration necessary to improve the mental health of the nation.

ECOLOGY OF MENTAL HEALTH MODELS*

By DONALD A. BLOCH, M.D.

> *"So little is known about what is in the Prime Minister's mind that had he complained that his talks in Washington had been a complete failure rather than a complete success, no one would have been the wiser."*[1]

MY PURPOSE in this chapter is to discuss some of the problems associated with the multiple construct systems which govern meaning in the mental health field. While psychiatry is not unique in having multiple and overlapping referential systems, it is, I believe, especially vulnerable to the charge that these systems and their activating cues are largely unknown to its practitioners. Basically this vulnerability rests on the unique place which science plays in legitimizing action programs in our field, it being only by this sanction that vast ventures such as the community mental health programs are permitted. Our ultimate ability to evaluate the effectiveness of these programs rests on our capacity to define the changes they are supposed to produce. To do this, I feel, we must undertake systematic, empirical study of the models or paradigms which define the values and guide the actions of that societal organ of which we are the parts: the mental health apparatus.

The term "model" as I am using it stands for an affective cognitive construct system. I shall not try to define it rigorously but mean by it a system of interrelated concepts and the terms, i.e., verbal symbols, which refer to them. It is generally understood that such models are built up slowly from infancy onward, that this process accelerates and jells with the acquisition of language, and that there is a simultaneous acquisition of parataxic and syntaxic definitions of verbal symbols. With development and acculturation these are hierarchically organized into systems of greater complexity, each of which has a configuration

*This paper is based on work done at the Jewish Board of Guardians in connection with NIMH Grant #MH 2446. Eva Rosenfeld is Co-Principal Investigator.

of its own, which may or may not be congruent with the substructures making it up.

A simple example from the physical world might be the familiar puzzle in which a wooden cube is assembled out of L-shaped elements. These latter are models, on their own level, as is the cube into which they are assembled, on its level. Gestalt psychology tells us much of how these matters are dealt with in regard to nonhuman events. It is an analogous study of such articulations and disarticulations with which we are concerned here.

In principle we might say that the mental health model exists whenever there is reference to it in order to give meaning to some experience of the person. Consider how this occurs with someone who will one day become a psychiatrist. Assuming he does not come from one of the growing number of hereditary psychiatric families, there is likely to be some identifiable point in his life when he has an experience and is told or reads that this is in the psychiatric domain. He is a changed man after this, possessing in embryo, for better or worse, a new referential system which will grow and subdivide. This process is, in general outline, similar for patient as well as professional. We say to a mother that her child's withdrawal is not healthy but pathological, that the enuresis is on psychological rather than urological grounds, that the temper tantrum is not badness but sickness—and thus, by a deft series of moves, carve out a new conceptual domain. We raise the unnoticed to awareness and reassign it and other experiences to the new model.

My point in starting with the primary, personal, experiential element in this process is to emphasize the continuing contribution of prototaxic, parataxic, and affective elements to cognitive model building, in the mental health field as well as elsewhere. From this point of view every construct system is unique and highly personal. Sections of it can be consensually validated to a greater or lesser degree, but never completely. Thus, every consensus is, in principle, somewhat false as well as somewhat true, the peculiar nature of language being such that our agreement about something may consist, at one extreme, of the matching of verbal symbols denoting completely different construct systems and, at the other of our matching, vast, subtle, and complexly ordered representations of our experience.

It is difficult to know how to mark off the construct systems which are specifically mental health in character. Nor is it easy to specify the cues evoking our reference to the mental health model. As psychiatrists, most of us have observed that there are times in listening to a patient when we assign the flow of similar material to quite different cognitive categories: that the patient ordered steak is a question of taste, that he will never drink milk may be quite another matter. The laws governing this choice of meanings seem self-evident to us when we are in their midst—but they are difficult indeed to specify.

The boundaries of the mental health domain are correspondingly indefinite. Biologists can say of mammals that they are warm-blooded vertebrates who have hair and nurse their young. We are at this disadvantage: if a lizard announced that it were a mammal, we could require it to prove that it nursed its young; there is no comparable test at present for construct systems which insist they relate to mental health. I think we might loosely define the category this way: it includes *events* which are psychological and interpersonal, the *purposes* of prophylaxis and healing and the *methods* of science for validation. The last, I believe, is critical in maintaining the current identity of the field. For example, religion and the law can be distinguished from mental health on this latter score, having recourse to revelation or tradition rather than experiment, as their principal methods of validation.

Whatever else may be said about these mental health models, they have shown an almost unbelievable vitality,* having in a short time captured for their referential terms much of the apparatus for social betterment, a great deal of that devoted to the education and acculturation of the young, to say nothing of the medical field. The change is "revolutionary" as Seeley noted in his address to the First International Congress of Social Psychiatry, revolutionary precisely because mental health constructs are used to justify political redistributions of goods, power, and symbols. (The part played by expert testimony in the integration ruling of the Supreme Court is the outstanding case in point. The existence of such enterprises as Mobilization for Youth, Har-

*As to "the reason why," Bertrand De Jouvenel, writing on "The Political Consequences of the Rise of Science," says: "In the history of European civilization it is easy to observe, first, a long era during which the men of thought were all men of God, clerics; then a gradual emergence of the men of law, who finally became the most favored and dominant type of intellectual. As we can tie great changes in political ideas and institutions to this displacement, we therefore have good reason to predict great changes from the supersession of the jurist by the scientist as the most favored and dominant type of intellectual."

—Bull. Atomic Scientists 3: 2-8, 1963

you Act, and the Social Psychiatry Congress itself, as well as the intense controversies raging over them, is also evidence of the invasion and its political consequences).

Thus, as far as the family of models goes, there is a population explosion: conditions to be treated—delinquency, multiproblem families, educational failure, indurated poverty, anomie, and community inaction are among those added to an already long list. It is an explosion of treatment modalities: therapeutic nursery schools, community mobilization programs, open and part-time hospitals, 24-hour emergency services, walk-in clinics. And of administrative units: NIMH, the Community Mental Health Boards, Regional Planning Committees, to mention a few.

Associated with these increases in number are major shifts in emphasis. As an example, consider the declining importance of psychoanalysis. Encomiums notwithstanding, this colloquium may, I think, properly be considered to be a mark of this decline. For example, Roy Grinker, in reviewing the history and development of training programs in psychiatry, says:

> "I refer particularly to our struggle for eclecticism in an effort to compensate for *our error* (my emphasis) in overemphasizing the principles and methods of psychoanalytic psychiatry. Contrary to the trend toward greater consideration of biological and pharmacological advances, it is the move from clinical studies of the patient-therapist dyad to the larger units (family, group, social, cultural, etc.) which has precipitated this struggle."[2]

Later in his article Grinker "reassures" his audience to the effect that:

> "accredited and senior analysts who accepted positions as directors of training in general psychiatry became transformed in that they assume responsibility for the training of general psychiatrists and not psychoanalysts."[3]

Thus, in the medical school the psychoanalytical model wanes in its influence, but this fact does not diminish the conviction of some psychoanalysts as to the importance of the medical model in their work. One group, commenting unfavorably on a recent proposal that certain lay analysts be admitted to full membership in the American Psychoanalytic Association, says unequivocally,

> "Medical and psychiatric training is more than an advantage to a psychoanalyst; it is the best foundation to the therapeutic practice of our profession as well as for scientific activities."[4]

These examples are not given to argue their merits but rather to illustrate the difficulty and complexity on this level of policy formulation, of identifying the appropriate construct systems to inform and give meaning to decisions; related to this is our inability to test, or know how to test, the extent to which actions have accomplished the desired ends.

A critical expression of this problem lies in our inability to carry on evaluative or outcome research. Assuming that we depend on the methods of science to validate and legitimize our treatment efforts, we must be able to demonstrate by these methods that *this* treatment causes changes whereas *that* treatment does not. Throngs of satisfied customers, as we know so well, do not testify to anything much *per se.* To show that they are testifying to something, we must be able to explicate the model of health to which they now conform. As a recent and painful illustration of the problems associated with evaluative research there is the experience of the *American Psychoanalytic Association* with its Central Fact Gathering Study. In his Presidential letter to the Association, Heinz Kohut speaks of the problems of a committee established to analyze these data collected some years in the past. He says:

> "In addition, we cannot deny that the work of this committee constitutes also a specific point of honor for psychoanalysts. As you know, the Association has been reprimanded by hostile critics from without and by stern censors from within who assert that we have committed the most heinous of all scientific sins, that of holding back the results of our research because they were inconvenient."[5]

Later on the same topic he says:

> The very shortcomings and disappointments of the fact-gathering research might become the seeds of further progress, and the very problems that have proved to be so disturbing (what is cure?, what is improvement?, what models can be used in assessing them?, what meaningful questions can be posed concerning 'the results' of psychoanalysts?) might well become the field for a thorough, empirically based investigation by a group of psychoanalysts working under the auspices of our Association."[6]

Presently the evaluative researcher struggles with an unknown number of mental health construct systems, of uncertain dimensional stability, which are activated by complexly contingent experiential cues. Small wonder he envies the physical scientist whose platinum bar, representing the standard linear yard, is nestled in its velvet case in the temperature- and humidity-controlled room in the basement of the Bureau of Standards.

We might improve our position by shifts in emphasis in current research. It is important to note, at the outset, that research has limitations, that science follows art, and that the work of model building and model explication goes on in many different ways. For example, the extensive Regional Planning process set in motion by the Joint Commission Report aims to discover the existing paradigms of illness and treatment by subjecting the issues to extended social argument. The seemingly endless sequences of subcommittees, advisory committees, task forces, executive committees, preliminary reports, and so forth, are not simply the props in an obsessive nightmare, nor are they exclusively action oriented. In the terms I am using, this is a vast machinery for consensually validating construct systems; it is also, of course, activity in the model building business.

Two broad lines of investigation suggest themselves: first, cognitive studies of individuals and, secondly, studies of the mental health apparatus using the techniques of cultural anthropology. In regard to the first, it seems reasonable to be curious as to the psychological function of a mental health value system for an individual, to wonder about its interpersonal consequences, and to consider how it relates to other characterologic or symptomatic expressions. Psychoanalysts would, I think, bring important skills to such research. Much relevant data should appear in the course of training analyses with special reference to the connection between personality, vocational choice, and developing career lines of young psychoanalysts, but this has rarely been studied systematically. Also as mental health activity increases there will be chances for sympathetic study of people who have had no previous contact with its ideas or language, as they first begin to do so. This might well provide us with an important opportunity to learn what its functional equivalents are, how it relates to other value systems, what it replaces, and with what consequences.

On the level of family studies it has frequently been observed that the official patient in the eyes of the family may be indistinguishable from the other family members as far as symptoms and other evidence of psychopathology are concerned. Grossly, we may guess that this is often a matter of blame laying or part of the current idiom for low rating another person. One group of investigators has suggested that the person identified by the family as ill has been chosen to *save* the family, providing us with an interesting parallel to the elements of omniscience and omnipotence observed by Sharaf and Levinson to influence career decisions in psychiatrists.[7]

At higher levels of analysis we might aim to track the shifting relations between politics, the law, religion, education, and the mental health mode, and hope to develop methods for testing hypotheses as to the reasons for, and consequences of, the vast expansion of our field.

Clearly there are far too many questions here, proposed at too many levels of discourse. They will give way only slowly, and it would be absurd to think of action programs waiting forever while careful studies are done to explicate their goals, the value premises underlying them, and the rules of inference on which judgments are based. But, since science legitimizes us (I am convinced that psychiatry's insistence on operating under a medical aegis reflects an intuitive appreciation of this sociological fact), we are obliged to engage in and give priority to inquiries directed at ourselves as creators and users of construct systems.

The research field is set out by Wallace, a cultural anthropologist, as follows:

> "The problem of description is not simply to translate an exotic nomenclature into English or some other, scientific language. In fact, exact translations can rarely be made. The problem is to define the taxonomic system itself—that is, to explicate the rules by which the users of the terms group various social and genealogical characteristics into concepts. It is a problem in cultural semantics, then, not in practical or structural linguistics, and as a semantic problem it is of cognitive and logical interest."[8]

Wallace is speaking of componential analysis of kinship relationships; our concern would be to discover the calculus underlying the making of mental health judgments. Such studies require a considerable and durable involvement on the part of diverse groups of clinicians and consumers of mental health products on one hand and researchers on the other hand. What are needed as data are the behaviors—i.e., clinical judgments, treatment decisions, dynamic formulations and so forth—that eventuate from contact of different individuals and different subgroups with similar clinical material. Hans Strupp's exposure of psychiatrists to the same filmed interview is an instance, as is Shirley Star's use of typical case histories in her interviews of a national probability sample. As Wallace puts it:

> "The commitment to describe the psychological reality of culture requires that not just any model which predicts some overt class of action be accepted, but only that model which is used as a system of reckoning by the actor."[9]

If we lend ourselves to such a study, it must not be to demonstrate that one construct system is more right than another, (this is a legitimate but different concern) but rather to see what they are and in what respects they differ. To accomplish this, the area of investigation itself must be highly valued by us all.

We have been fortunate at the Jewish Board of Guardians to work with clinicians on this basis, but much hard work and good will must go into any such joint clinical-research venture, and many end disastrously. Indeed, psychoanalysts could make a major contribution if they could help us to achieve better dynamic understanding of the frequent hostility between clinicians and researchers. The problem is serious, ubiquitous and seems to me to be a clear expression that differing calculi are being used, although each group professes the same goals of advancing knowledge.

In the title of this chapter, it has seemed useful to me to use the metaphor of ecology—". . . the branch of biology which deals with the mutual relations among organisms and between them and their environments"[10]—to underscore the point that there are a large and unknown number of mental health construct systems and that they exist in dynamic interaction with each other and with such other families of construct systems as the religious models, the political models, the physical health models, and so on. One may almost think of them as if they were a plant or animal family with many species and subspecies, dwelling in certain habitats, waxing or waning in numbers and variety according to circumstance, changing the environment, and being changed by it.

The particular appeal of this metaphor for me is that it presupposes taxonomy—psychiatrists work endlessly at classifying their patients, but very little at classifying the construct systems whereby they classify their patients. In addition, ecology speaks of the laws of change and succession in which there is "the gradual replacement of a . . . species on a given site and species which invade originally are unable to continue when the habitat becomes more densely settled (and) a community 'spoils' its own habitat."[11] Somehow these considerations seem especially compelling in this time of the success of our models.

Robert Merton has said:

"It is true that, logically, to establish the empirical genesis of values and beliefs is not to deny their validity, but this is often the psychological effect on the naive mind."[12]

And, in the same volume:

> "There is ample institutional evidence of this failure to follow up through empirical research the numerous and now widely recognized problems of the relations between science and the social structure; nowhere among the universities of this country is there an Institute for Research on the Social Relations of Science."[13]

Anne Roe has said:

> "The artist exposes himself in his work; the scientist seems rather to hide in his, but he is there."[14]

As psychoanalysts we have been uniquely willing to agree with this sentiment and to regard it as relevant to the work we do. Perhaps it is now our responsibility to create the analogue, appropriate to this age, of the requirement that the individual analyst be analyzed. This is to try to study ourselves as cultural products, to see our work as embedded in the broad flow of historical and social process—and to see it that way in detail and as a matter of practice, not simply as a matter of faith.

REFERENCES

1. Worsthorne, P.: quoted by John Crosby. New York Herald-Tribune, December 16, 1964.
2. Grinker, R.: A struggle for eclecticism. Am. J. Psychiat. 121: 451-457, 1964.
3. *Ibid.*
4. Karush, A., and Shapiro, D.: Supplementary Statement of the Council of the Association for Psychoanalytic Medicine.
5. Kohut, H.: Newsletter. Am. Psychoanalyt. A. 5:1964.
6. *Ibid.*
7. Sharaf, and Levinson: On the quest for omnipotence in professional training. Psychiatry 27:135-149, 1964.
8. Wallace, A.: Culture and cognition. Science 135:351-356, 1962.
9. *Ibid.*
10. Webster's New International Dictionary, 2nd ed. Springfield, Merriam, 1934.
11. Newman, J. R., ed.: Harper Encyclopedia of Science. Washington, Harper and Row, 1963.
12. Merton, R. K.: Social Theory and Social Structure, Glencoe, The Free Press, revised ed. 1957, p. 547.
13. *Ibid.* p. 533.
14. Roe, Anne: The psychology of the scientist. Science 134:18, 1961.

THE HUMAN PREDICAMENTS OF AN ISLAND POPULATION

By MILTON MAZER, M.D.*

THIS VOLUME is responsive, I believe, to a major trend in American psychiatry, the increasing interest in social psychiatry and its application, community psychiatry. The trend is based on the recognition that the human personality is not completely shaped by the age of six, and is not impermeable to the vicissitudes of life, thereafter. Whatever the psychological substrata which supports an episode of mental disorder, the occurrence of the disorder at a particular moment in the life course of the individual raises immediate questions which are important for theory and vital for practice.

In addition to the traditional concern with the basic personality the therapist needs to be equally concerned with the circumstance, interpersonal or social, which causes a decompensation in the accustomed psychic equilibrium of the individual. Among work which has focused our attention on such considerations is that of Lindemann[1] on bereavement and related crises, and that of Grinker and Speigel[2] on the stress reactions associated with military combat.

In addition to such considerations the trend toward a community psychiatry has been accelerated by two other developments. The first of these is based on work showing the intimate relationship between mental disorder and the social, cultural and economic circumstances under which people live. In short, we are coming to realize that whether mediated by parent, by sibling, or by other persons, mental disorder is insistently social in origin. The second force to which I refer is our new knowledge of the high frequency of mental disorder with an increasing demand for treatment facilities. This tremendous demand

*I am indebted to the following physicians of the Island for their collaboration in the general practice phase of the study: Joseph Frisch, M.D., Russell S. Hoxsie, M.D., Donald R. Mills, M.D., Robert W. Nevin, M.D., and David Rappaport, M.D.

This investigation was supported by Contract PH 43-64-493, Biometrics Branch, National Institute of Mental Health.

makes the intensive, long-term treatment of a minority by a significant proportion of our superiorly trained therapists unresponsive to the needs which confront psychiatry as a profession.

We are further becoming aware that not all or, perhaps, not even a majority of those with mental disorders are seen by the mental health professions. As someone has put it, a psychiatric patient is "someone who complains or is complained about, and these two categories include the neurotic and the psychotic patients, respectively. On the other hand, there are large groups of persons who neither complain nor are complained about in this personal sense, but who experience a variety of human predicaments which may well be termed parapsychiatric. Examples of these are desertion, divorce, premarital pregnancy, criminal actions, juvenile delinquency, repeated single-car automobile accidents, educational difficulties and chronic alcoholism.

The population under study in this project is that of the Island of Martha's Vineyard lying 7 miles by sea off the southeast coast of Massachusetts. The Island, roughly triangular in shape, consists of low hills along its two northerly sides with a plain dotted by fresh water ponds extending to its southern shore. The Island is 106 square miles in area, approximately five times that of Manhattan. Its population density, however, is quite low, averaging fifty-five persons per square mile, only slightly higher than the average for the United States as a whole. Although part of the great urban coastal region extending from southern New Hampshire to northern Virginia which Gottmann[3] has called Megalopolis, it remains one of its most sparsely settled areas.

The Island of Martha's Vineyard, and the neighboring Elizabeth Islands, make up Dukes County. As a result, there is a wealth of data concerning it in the United States Census reports. Since the neighboring Elizabeth Islands had a population of but sixty-six in 1960, the data for Dukes County in the census reports may be used for that of Martha's Vineyard with but negligible error. The population of the County in 1960 was 5,829 persons. It has been remarkably stable in numbers, increasing only 4.6 per cent per decade during the 170 years since the first census of 1790. During the ten years, 1950 to 1960, the population increased by 3.5 per cent as compared to 9.8 per cent for Massachusetts as a whole.

The population is almost entirely rural, nonfarm; 85 per cent of it lives in the three down-island towns of Tisbury, Oak Bluffs, and Edgartown, with the remaining 15 per cent sparsely settled over the half of the Island known as up-island. The major occupations are dependent upon the Island's principal industry, the summer tourist trade, and involve services and housebuilding and repairing. Fishing and farming, once significant enterprises, are rapidly diminishing in importance. The two large ethnic groups are "Yankee" and Portuguese. The three small, nonwhite racial groups consist of Indians, Negroes, and descendants of Cape Verde Islanders, the so-called Portuguese 'bravas.' The nonnative born population arises from a number of countries, the two largest groups coming from the Portuguese Islands, the Azores, Madeira, and the Cape Verde Islands, and from the Maritime Provinces of Canada. The age of the population is heavily weighted toward the higher end of the age scale; 17.4 per cent are over 65 years of age as compared with 11.1 per cent for Massachusetts. There is a relative deficiency of persons in the posthigh school age group due to the out-migration of a large proportion of those finishing high school.

The Island is a low income area, having the lowest median family income of any of the fourteen counties of Massachusetts. The median family income in 1959 was $4,745 as compared with $6,272 for Massachusetts as a whole.

The study here reported grew out of the establishment of a community mental health program beginning in the summer of 1961. It was soon clear that a unique opportunity presented itself. Because of the manageable size of the population, an unduplicated count of those consulting psychiatric facilities could be secured. Because migration was minimal, the population at risk could be readily determined. Because of the informality and intimacy of caretaker relationships, it was soon clear that a great deal of demographic data could be secured.

The psychiatric register was adopted as a technical model. These have been established in recent years for a number of communities in order to provide for a prospective, continuous recording of mental disorders. They vary in the size of population covered from 31,000 in a health area in Bronx County, New York, to 3,000,000 persons in the State of Maryland.[4] The registers have been designed to provide: (1) unduplicated counts of persons in an area receiving psychiatric care, (2) an evaluation of the long-term outcome of psychiatric care programs, (3) to note changes in diagnoses over time, (4) to throw light on the natural history of mental disorders, (5) to study patient flow among facilities, and (6) to determine the chronicity of psychiatric care.

Gorwitz and his associates have reported on the maintenance of the register for the State of Maryland.[5] Gardner and his associates have reported on the first year's experience of the register for Monroe County, New York.[6] The maintenance of a register for a small parish in Scotland for one year has been reported by its psychiatrically trained general practitioner.[7]

It is clear, however, that a psychiatric register can record psychiatric illness only in those who are aware of its presence or in those instances in which the illness is sufficiently severe to require intervention by others as in some cases of psychoses. There are, however, categories of behavior generally called "acting out" which are presumably related to psychological disorder. Instead of the impairment coming within the awareness of the individual, he finds himself in difficulty in his relationships and in his social integrations. The cases of those whose disorders are expressed in socially visible acting out rather than in awareness are usually unrecorded in psychological and psychiatric records. Consequently, counts made by psychiatric registers alone may be expected to underestimate the frequency of interpersonal or individual-societal maladaptation in a population. Further, if awareness of distress and the decision to seek expert help are a function of social class position or ethnic origin, limiting the register to treated psychiatric disorders will give a biased picture of the frequency of disorder in each social or ethnic group.

The present paper, which is a preliminary report of a study in progress, suggests a means for recording, in addition to treated psychiatric disorders, the major groups of human predicaments which are presumed to be signs of interpersonal and individual-societal maladaptation. Two early epidemiological studies of mental disorder by Rosanoff[8] in Nassau County, New York, and by Roth and Luton[9] in Williamson County, Tennessee, included data on a number of human predicaments which are not self-defined as psychiatric. An elaborate study of such predicaments was conducted in the suburban community of Wellesley, Massachusetts by Devitt and her associates[10] during the month of March 1953. They secured from community caretakers the names of and demographic data about those who were suffering from three categories of disturbance; dependency, ill health and psychosocial maladjustment. The rate for diagnosed mental illness was found to be 4.5 per thousand persons and for maladjustment in the psychosocial area 10.3 per thousand, giving an overall rate of those conditions of 14.8 per thousand persons. In addition, Devitt and her associates related the occurrence of the predicaments studied to sex, age and ecological areas of the community.

A similar investigation limited to the mental health area has been conducted in the same locale by Kelly, Simmons and Baldwin[11] during the month of April, 1959. One result of the study was the recommendation that such reporting be done continuously by a method built into the community's social welfare structure.

After preliminary survey and discussions with possible sources of data in the community, a psychiatric—parapsychiatric register was established on March 1, 1964 for the population of Martha's Vineyard. For the purpose of the register, a psychiatric case was defined as a person in treatment with a psychiatrist, hospitalized in a mental hospital or in treatment with a physician for a psychiatric condition. A psychiatric condition is one of the disorders described in the Diagnostic and Statistical Manual of the American Psychiatric Association published in 1952. A parapsychiatric case was defined as a human event which is presumed to be the behavioral manifestation of interpersonal or individual societal maladaptation and which is socially visible and/or in the public records. The following are the list of parapsychiatric predicaments which were included in the register; chronic alcoholism, nonmarital pregnancy, suicide and suicide attempt, marital desertion, separate support and divorce, criminal convictions, the establishment of juvenile delinquency, clearly culpable auto accidents (single car auto accidents), major school disciplinary problem in the high school, psychological counseling in the high school, and acute alcoholism requiring temporary jailing. Welfare recipiency was included because of a consideration to be noted later.

Every psychiatric and parapsychiatric experience occurring to an Islander is recorded in the register along with whatever demographic data can be secured. In the case of psychiatric patients, virtually every item of demographic data available for counties in the United States Census of Population is secured for each person. The other data are secured from the usual sources including the District Court, the school system, the Registrar of Motor Vehicles, the Departments of Welfare, and the like. For the three month period March 1, 1964 to May 31, 1964 five of the six general practitioners completed a card for each of their patients presenting a significant psychiatric disturbance. Each card was reviewed briefly at monthly intervals in a meeting with the psychiatrist in order to establish a concensus as to the terminology used. In order to study possible familial clustering of persons with predicaments, it is necessary to secure data on Island families. This is to be done by utilizing the data secured by a door-to-door religious census conducted by the Island's churches in 1962; at present the project has family cards on approximately 75 per cent of the population. Changes in the population at risk by migration are to be determined by comparison of the yearly voter registration census and by other reports. The purposes of the register are as follows: (1) to provide an unduplicated count of human predicaments in a relatively stable population, (2) to determine whether the self-defined psychiatric predicaments and those subsumed under the heading "acting out" (parapsychiatric events) come from the same or different populations, (3) to study the familial clustering of human predicaments in order to identify high incidence families and by comparing such families with those of low or no incidence to determine which factors are associated with multiproblem families, and (4) finally, it is hoped that the identification of high incidence families may result in the development of a program of anticipatory intervention.

The Prevalence of Treated Psychiatric Disorders

The mental health expert planning a psychiatric program for a community requires data on the probable number of persons in the community requiring psychiatric care. Borrowing from public health practice mental health experts have often used prevalence and incidence rates. By prevalence is meant the number of persons in the community suffering from a particular disease at a particular point in time or through a defined period of time. If the point is an instant in time, or more frequently, a particular day of the year the rate is called a point prevalence rate. If the period is a longer interval such as three, six or twelve months, the rate is a period prevalence rate. By incidence is meant the number of new cases of a disease which appear during a defined period of time. Incidence rates are particularly useful in acute illnesses or in illnesses whose point of origin in time can be readily determined. Since this is generally not true for psychiatric illness such incidence rates as are used are usually computed on the basis of the time when the person entered a treatment facility, either outpatient or hospital.

The prevalence of treated psychiatric illness both on an ambulatory basis and in a mental hospital is given in table 1. If a patient was seen for three, six and twelve month periods as an outpatient and also hospitalized, he is recorded in the hospital column. The number of cases given may be considered to approximate the total number of Islanders in psychiatric treatment. During the periods of study only three Islanders were known to have been treated by psychiatrists off the Island. Two of these were also treated in the Island's mental health facility; all three are included in the table. Patients who are hospitalized in hospitals not usually used for Island admissions may have been missed. But if there were any in this category, the number is small.

TABLE 1.—*Number of Patients and Period Prevalence Rates*

	Number of Patients			Rate Per
	Outpatient	Hospital	Both	1000
3 mos. (3/1/64 to 5/31/64)	51	18	69	11.8
6 mos. (3/1/64 to 8/31/64)	58	23	81	13.9
12 mos. (9/1/63 to 8/31/64)	78	31	109	18.7

The table shows that the prevalence rates for treated psychiatric disorder in the community were 11.8, 13.9, and 18.7 for the three-, six-, and twelve-month periods. It must be emphasized that these rates are neither sex nor age adjusted and, therefore, cannot in the form presented be used for accurate comparisons with rates from other communities.

DIAGNOSTIC CATEGORIES AND TYPE OF CARE

It is well-known that most patients with psychiatric symptoms are treated by the general practitioner and the present study shows that in the Island population this is true of 79.5 per cent of defined psychiatric illnesses. Diagnostic categories into which our psychiatric patients fell by type of care received is shown in table 2. Those listed under the column marked "Psychiatrist" include those seen both on an ambulatory basis and hospitalized in a mental hospital. The final column is a summation of those seen by the psychiatrist and by general practitioners. The data in this column show that during the three month period of this prevalence study, the largest number of patients fell in the category "Psychoneurotic Disorders."

They made up 48.7 per cent of the total defined psychiatric patient load. The next largest category was that of "Psychophysiologic Disorders" making up 26.1 per cent of the total patient load. The same

TABLE 2.—*Distribution of Patient Load by Diagnoses*

	Psychiatrist		Gen. Pract.		Both	
	No.	%	No.	%	No.	%
Organic brain syndrome	2	2.9	17	6.3	19	5.6
Psychotic disorders	21	30.3	7	2.6	28	8.2
Psychophysiologic disorders	5	7.2	84	30.8	89	26.1
Psychoneurotic disorders	20	29.0	146	53.7	166	48.7
Transient situational personality disorders	14	20.3	11	4.0	25	7.3
Others*	7	10.3	7	2.6	14	4.1
Total	69	100.0	272	100.0	341	100.0

*Includes personality disturbances, special symptom reactions, and mental deficiency.

ranking of categories was true for the general practitioner. The experience of psychiatric facilities, however, was somewhat different. The psychotic and psychoneurotic categories were the two largest with the transient situational personality disorders, largely disorders of chil-

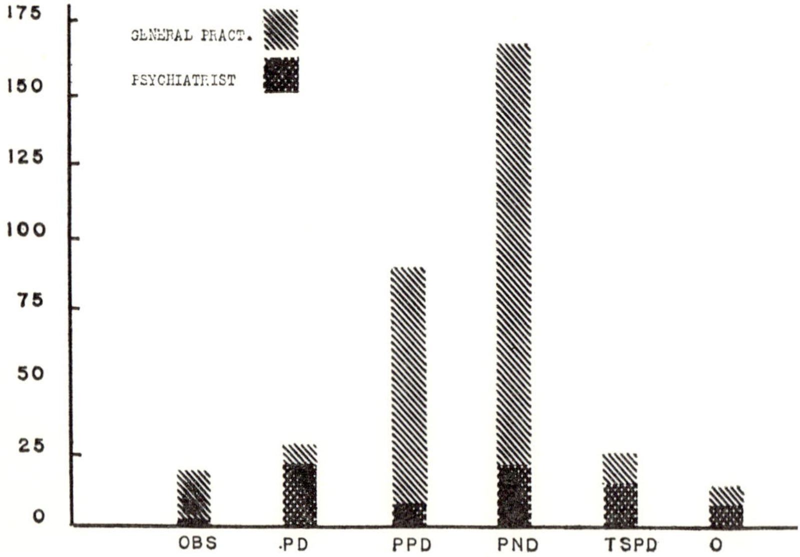

FIG. 1.—Number of patients by type of care. OBS: Organic Brain Syndromes PD: Psychotic Disorders PPD: Psychophysiologic Disorders PND: Psychoneurotic Disorders TSPD: Transient Situational Personality Disorders O: Others (Personality Disorders, Special Symptom Reactions, Mental Deficiency).

dren, in third place. The data of table 2 is shown graphically in figure 1. The apparent elevation in the center and decline toward both ends of the graph indicated by the bars has, of course, no significance. It is simply an accident of arrangement of the diagnostic categories.

The data also provide an answer to the following question: are there biases among diagnostic categories for the type of professional service secured? The data are shown graphically on figure 2. Approximately 90 per cent of patients with organic brain syndromes, psychophysiologic disorders and psychoneurotic disorders were treated by general practitioners. In only the psychotic disorders group were a preponderance of patients treated by a psychiatrist. These data have certain obvious implications for psychiatric educational programs for medical students and general practitioners.

The Prevalence of Human Predicaments

For the purpose of this study, a human predicament is any psychiatric or parapsychiatric event as defined in the introduction. The number of persons experiencing human predicaments for a three-month

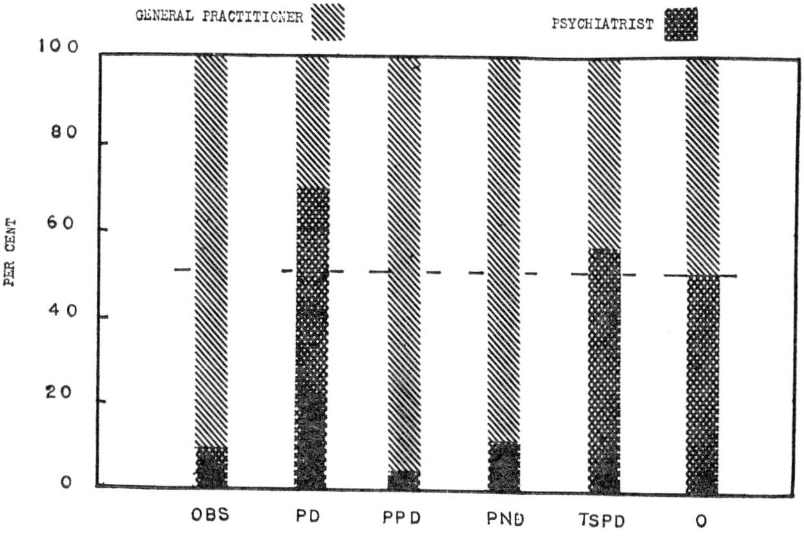

Fig. 2.—Per cent of patients by type of care. OBS: Organic Brain Syndromes PD: Psychotic Disorders PPD: Psychophysiologic Disorders PND: Psychoneurotic Disorders TSPD: Transient Situational Personality Disorders O: Others (Personality Disorders, Special Symptom Reactions, Mental Deficiency).

period is given in table 3. If a patient was seen by both the psychiatrist and a general practitioner, he is counted only in the psychiatrist column. In the case of other multiple recordings a psychiatric event takes precedence in recording over a parapsychiatric event or welfare recipiency, and a para-psychiatric event over welfare recipiency. During the three month period there were 69 persons in treatment with a psychiatrist either in a hospital or on an ambulatory basis. Two hundred sixty-nine persons were in treatment for a significant psychiatric disorder by a family physician. The data show that only a fraction of persons in the community suffering from psychiatric disorder were treated by the psychiatrist, even though the patient-psychiatrist ratio in the community is similar to that of many urban centers, namely, 5,829 persons per psychiatrist. To put it another way, 79.5 per cent of patients suffering from significant psychiatric symptoms were treated by general practitioners. The treatment rate by a psychiatric facility was 11.8 per thousand persons and by general practitioners 46.1 per thousand. The combined rate was 57.9 per thousand which indicates that 5.8 per cent of the population was in treatment for a psychiatric condition during the three-month period of the study. In addition there were 90 persons who came into the register as a result of experiencing one or more parapsychiatric events. The parapsychiatric subject rate was 15.4. This, when added to the total psychiatric rate, gives a rate of 73.3 per thousand persons which indicates that more than 7.3 per cent of the population suffered from some human predicament with psychiatric implication during the three-month period of this substudy.

Welfare recipiency is noted in the table because a number of studies have shown it to be highly correlated with problems of emotional dependency and maladjustment. It is not suggested that welfare recip-

TABLE 3.—*Psychiatric Disorders, Parapsychiatric Disorders, and Welfare Recipiency (3-Month Prevalence)*

	Number of Persons	Rate per 1000	Cumulative Rate
Psychiatric patients	69	11.8	
General practice patients	*269*	46.1	
Total psychiatric patients	338		57.9
Parapsychiatric subjects	90	15.4	73.3
Welfare recipients	144	24.7	98.0

iency has the same import as the other human predicaments. Many instances may simply be the result of advanced age and the economic vicissitudes of life. Nonetheless, in our Calvinistic society, the need to accept the bounty of one's neighbors is likely to be productive of emotional stress. Further analysis will determine the degree of correlation between welfare recipiency and the other human predicaments.

Summary

The paper has described a technique for a prospective, ongoing recording of a wide range of human predicaments having psychiatric implications. The method used is an elaboration of the psychiatric register.

Period prevalence rates for psychiatric disorders in treatment by psychiatrists and general practitioners and for a wide range of para-psychiatric events are given for a relatively confined low income, rural population. The data show that over a three month period 5.8 per cent of the population was in treatment for a psychiatric condition with either a psychiatrist or a general practitioner. When the para-psychiatric subject rate was added to the psychiatric subject rate it was found that 7.3 per cent of the population experienced a human predicament with psychiatric implications during a three month period.

Even though the community had a psychiatrist-patient ratio of one to 5,829, comparable to that of most urban centers, approximately 80 percent of patients with significant psychiatric disorder were treated by the general practitioners of the community. It is evident that the actual treated case rate is beyond the psychiatric resources of the community under study or, for that matter, of any existing American community. This finding supports the desirability of extensive training of non-psychiatric mental health practitioners.

The tabulation of the psychiatric case load by categories and type of service showed that approximately 90 per cent of persons with organic brain syndromes, psychoneurotic, and psychophysiologic disorders were treated by general practitioners. The psychophysiologic disorders reached a psychiatric facility infrequently. In the case of the psychoses alone did the psychiatric facilities see a preponderance of the patients. These data have clear implications for the design of training for medical students and general practitioners.

In conclusion, I should like to emphasize the need for further studies based on methods which will identify instances of psychiatric disorder which are so defined neither by the patient nor by his relatives and which reach the attention of society in the records of the courts, the schools, Registrar of Motor Vehicles and social and other agencies. The data presented are to be taken as preliminary and illustrative and by no means a sufficient description of the mental health problems of even so small a community as Martha's Vineyard.

REFERENCES

1. Lindemann, E.: The Wellesley Project for the study of certain problems in community mental health. *In* Interrelations between the Social Environment and Psychiatric Disorders, New York, Milbank Memorial Fund, 1953, pp. 165-185.
2. Grinker, R. R., and Spiegel, J. P.: Men Under Stress. New York, Blakiston, 1945.
3. Gottmann, Jean: Megalopolis. Cambridge, The M.I.T. Press.
4. Bahn, Anita K.: Psychiatric Case Register Conference, 1962, Public Health Reports 77:1071-1076, 1962.
5. Gorwitz, K., Bahn, Anita K., Chandler, Caroline A., and Martin, W. A.: Planned uses of a statewide psychiatric register for aiding mental health in the community. Am. J. Orthopsychiatry, 33:494, 1963.
6. Gardner, E. A., Miles, H. C.; Bahn, Anita K., and Romano, J.: All psychiatric experience in a community. Arch. Gen. Psych. 9:81, 1963.
7. Primrose, E. J. R.: Psychological Illness: A Community Study. Springfield, Ill., Charles C Thomas, 1962.
8. Rosanoff, A. J.: Survey of Mental Disorders in Nassau County, New York, July—October, 1916, Psychiat. Bull. 2:109, 1917.
9. Roth, W. F., and Luton, F. H.: The mental health program in Tennessee. Am. J. Psychiat. 99:662-675, 1943.
10. Devitt, M. *et al.*: The Predicament Study. Wellesley, Mass., The Human Relations Service of Wellesley (mimeo.).
11. Kelly, J., Simmons, A. J., Balwin, J.: Mental health situations in a Boston suburb: A method for community reporting on mental health problems. *In* Suburban Ecology and Impaired Behavior: Examples of Community Mental Health Research, A. J. Simmons, ed. In preparation.

CONTRIBUTIONS OF PSYCHOANALYSIS TO SOCIAL PSYCHIATRY

By LEONARD J. DUHL, M.D.*
AND
ROBERT L. LEOPOLD, M.D.

IN THE last several decades, the boundaries between eclectic psychiatry and psychoanalysis have become somewhat indistinct, and ideas have flowed in both directions. Similarly, the lines dividing general psychiatry (whether psychoanalytically oriented or not) from social psychiatry have tended to blur. Let us therefore look at some of these recent developments.

Social psychiatric concepts have evolved gradually from those long recognized by psychiatry and its allied professions. The concern of the modern therapist with the social context of the patient has been equated by some with social psychiatry. But, for the purpose of this chapter, we trace a wider arc. Our concern is with psychiatric interventions in which the predominant relationship is not with a patient, but with people who have a critical effect on agencies, institutions, or community organizations. Such interventions most usually occur outside of the traditional psychiatric milieu.

It must be stressed at the outset that everything a patient does and says, including what he does or says as a participant in a social system, falls within the therapist's purview. Thus, every psychiatrist and every psychoanalyst who maintains the fundamental one-to-one relationship of individual treatment must equally maintain an indirect relationship to the patient's social system. Indeed, following the work of Nathan Ackerman in dealing with the total family, it is getting to be more common to have direct contact with the patient's family and other significant figures in his environment. A modern therapist, then,

*We wish to thank Mrs. Dorothy Kuhn, research assistant to Dr. Robert Leopold, Mrs. Antonia Handler Chayes, Program Planning Specialist, NIMH, and Professor John R. Seeley, Ford Center for Behavioral Sciences, Palo Alto, California.

although he may not practice social psychiatry in the full sense delineated here, does in fact engage in some aspects of it.

In the twentieth century new confrontations with social systems have created new kinds of relationships among psychiatrists, individuals and groups. The experiences of three wars in this century erected the signposts of a new way of working. In the course of these wars, the psychiatrist was asked to lend his special knowledge and insight to military programs of selection, training, support in the field, and rehabilitation. The psychiatrist soon found he could not depend exclusively or largely on classic one-to-one relationships with individuals to meet the mental health needs of the military services. He had to become increasingly involved in more complex interactions with many kinds of military and quasi-military personnel, i.e., with the whole social system serving the war effort. Just as he had traditionally sought to intervene in the individual's functioning in the interest of mental health, so he now sought to intervene in the functioning of a social system to reach the same ends.

The experience of the psychiatrist in new situations created by these emergencies and met by improvisation gave him a foretaste of what it would mean to take care of the psychological needs of a total population since once again traditional practice would have to be modified in order to meet the vast demand with inadequate manpower. At the same time, social scientists began to see how psychiatry might contribute to the search for an answer to a question which had been brought into sharp focus by the war: "What is wrong with a society that produces so many 'misfits?' " Thus, if the control of certain environmental factors affected the rate of incapacitating mental illness in a military situation, inevitably the possibility arose of intervening, with the same objective, in nonmilitary situations. And so the concept spread to other socal institutions; for example, industrial organizations, schools and colleges, philanthropic overseas service agencies, and more recently, the Peace Corps.[1]

Interest in the concept also spread to clinical or medical environments. In 1954 Stanton and Schwartz[2] published their classic study on "hospitalisms"—the hospital's role in fostering mental disease. Their findings were substantiated by William Caudill[3] and others. These studies indicated that by managing the hospital's own *social system* it could be made to serve a significant therapeutic purpose instead of being destructive.

Today the psychiatrist's participation in such societal interventions is more or less taken for granted. For example, a university psychiatrist may practice not only individual psychotherapy, but may also be deeply involved in the university's social system, where he will be trying to develop a sound mental health climate. His presence at administrative meetings, and even classes, may seem an unorthodox form of psychiatric intervention, but its roots in traditional psychiatry and its relevance can be demonstrated.

Social psychiatry therefore begins to show concerns common to those of public health. In addition to the interest in care, treatment and rehabilitation, there is concern with all the environmental factors relevant to the promotion of health. And so the emphasis shifts from curing the sick to improving mental health in the whole community.

Social psychiatrists, like public health physicians, find themselves working in state and local health departments, educational systems, welfare bureaus, and many other governmental and quasi-governmental agencies which in one way or another affect the mental health of significant numbers of people. The psychiatrist's predominant role is that of a consultant to those directly affecting a social system; for example, as a consultant to principal or teachers in a public school. He will also intervene indirectly in the functioning of the whole system in order to promote the mental health of the population for whom the agency is "responsible." An increase in the number of patient referrals from one part of a large program often serves as a signal of some serious disturbance in the organization. By identifying a problem which leads to a more careful study of administrative procedure and policy by responsible officials, a psychiatric consultant can often have an important impact not only on preventing breakdown, but in improving the general conditions of work and promoting productivity.

Where concern with the social system as a whole began with improvised responses to emergency needs, continued involvement has demonstrated the need for thinking ahead and planning for environments conducive to mental health. Inevitably, such planning requires participation in the overall policy and program for the institution or community.

If one accepts, *arguendo,* that the psychiatrist has a significant professional role in a social system, it becomes important to explore more closely the contributions that are offered by psychoanalysis.

Psychoanalytic Bridges to Social Psychiatry

How can people who have specialized in exploring individual depths function effectively when they move into activities primarily concerned with social systems? There are difficulties, to be sure—groups of individuals and institutions do not "behave" in the same manner as a single human being. However, psychoanalytic theory has important insights to combine with the understanding and theories of the other social sciences whose major concern is focused on institutions.

Psychoanalytic theory has played a part in the formulation of some important elements of what is called a general systems model of society. And psychoanalysis provides, in itself, a general systems model. Such a model places the individual in a vast and interacting network of systems. He himself is a system—each of his sub-systems (vascular, respiratory, muscular, digestive, etc.) involves constant and reciprocal interchange of matter and energy. The individual as a system is in a similar process of interchange with a series of impinging systems. Starting with his immediate family, the network branches out to include systems represented by all the "communities" in which he lives: educational, occupational, religious, political, recreational, geographic, etc. These systems interact with each other both directly and indirectly. Every part of this tangled net of systems touches the individual in a multitude of ways, and he himself affects every part of the net. To affect one part of the complex set of systems is to set in motion a series of reactions in others. Psychoanalysis as a therapeutic technique deals with the individual and by so doing sets in motion many reactions in other parts of the individual's various social systems. Likewise, he is aware that changes in other parts of the systems network (e.g. job changes, education) may set in motion reactions which may in some cases be nearly as significant to the individual as the psychoanalytical process itself.

Thus, even the psychoanalyst who focuses on the individual has a sense of process with respect to the interactions between the individual and the ever-widening social systems of which he is a part. This sense of process—this awareness of the complex functioning of the social system—is indispensable to the social psychiatrist. Psychoanalysis was one of the first branches of modern medicine to deal with multiple casualties. It recognized no single cause for mental disorder, neither in the individual, nor in the environment. Rather, illness was the product of a complex combination and interaction of disturbances within the individual and between him and his environment. Because he feels comfortable with the concept of a general systems model, the psychoanalyst can bring to the problems of mental health and illness an appreciation of the complexity of interrelationships between the individual and the social contexts in which he functions.

The conceptual framework of individual psychoanalytic psychotherapy provides some significant *bridges* to the practice of social psychiatry. In community-wide preventive efforts, social psychiatry is concerned with sociocultural supports required for individual psychological health, especially at critical stages of personality development.

One example of particular urgency at the moment is the weakening or disappearance of institutions that formerly helped the adolescent find his ego identity. Involved here is the dissolution of the strong family unit, the ever-growing multiplicity of functional "communities" of which the individual is a part, the political and other jurisdictional complexities which blur group identities, the increasing difficulties involved in making a career choice, the gaps between the education generally available and the skills needed to survive, and other problems all too familiar to most students of human affairs. The psychiatrist who is called on to assist in planning sociocultural institutions that, hopefully, will compensate for these deficiencies in our social life, can be expected to plan more effectively if he is equipped with the psychoanalyst's special orientation to the concepts of *levels of ego growth and of socialization.*

Recent research concerned with the impact of the social environment on human behavior points to the close relationships of psychoanalytic and sociological theory. Fried[4] in the West End study, proposes a principle of "psychosocial complementarity."

> He says: "Adaptive behavior is based on the central regulative function of internal resources and controls when mechanisms of social regulation are relatively undifferentiated and role definitions are diffuse and flexible; and it is based on the central regulative function of external resources and controls when mechanisms of ego regulation are relatively undifferentiated and goal directions are ambiguous and limited."

Such developments in psychoanalytic and social theory unerringly point to both further research and to extended activity of analysts in dealing with social systems and community problems. In such undertakings the skills of psychoanalysis alone will not answer the many questions—for the field of social psychiatry is not the sole preserve of psychoanalysis.

Another bridge is the analyst's sensitivity to the *timing of interventions.* The social psychiatrist who seeks to promote the cause of mental health views it not only in individual terms, but also in community terms, particularly with respect to those community resources which support the individual when his internal resources are inadequate for coping with stress. For the psychiatrist, a thorough understanding of crisis theory is indispensable. He must be acutely aware that crises occur in social systems as they do in individuals, and what the significant differences are.

Certainly an institutional crisis involves people; yet frequently it relates to such problems as administrative structure, legal entanglements, financing and capitalization. Crises such as these are not the prime concern of the social psychiatrist, but one he should be cognizant of in the process of consultation with key individuals. He must be aware that personal crises in the individuals who most affect the functioning of the institution may reverberate throughout the whole organization.

For the social system, as well as for the individual, a crisis may present an opportunity for growth and maturation as well as a risk of stagnation. The psychiatrist must know, furthermore, that the social system in crisis, like the individual in crisis, is more susceptible to influence than at other times of life. Thus, he makes the most of crises, using them to intervene in the social system in the interest of bettering its resources for supporting and promoting mental health.

> An organization which had consistently been successful in all its operations was faced with increased incidence of illness and accidents, low production, decrease in morale, and breakdowns in formal and informal communications, and was ready to discharge several key persons in order to "solve" the problem. The psychiatrist, by being alert to the "flags" of disturbance, was able to study the total situation and by redefining the problem for the responsible administrator enabled him to find alternative solutions and resolution of the crisis. Though the institutional crisis manifested itself in part with problems of individuals, the solutions, found with the help of the consultant, at a time when the organization was in great stress, led to administrative and institutional changes in procedure which radically improved the situation. To have noted the same problems when the organization was not feeling stress would have led to no action. Here the sense of timing in institutional settings is critical.

A psychiatrist may know the dynamic structure of a patient's problems many months in advance of his interpretation to the patient. But he must learn how not to intervene too quickly in order to avoid a direct attack on the patient's defenses. Teaching this abstention is a conspicuous difficulty in training young psychiatrists. In social psychiatry, it is equally important that the consultant withhold intervention, even though he is sure he understands a problem and can help find an answer to it, until the social system is ready to loosen the defenses it has built up against seeking a solution. But understanding the timing of interventions is even more difficult for the social psychiatrist, since there is no established body of experience.

The social psychiatrist, to guard against too early intervention, must understand thoroughly the system he is working with. He must also understand himself in order to avoid intervention timed in the interest of his own needs. Previous individual practice offers experience in timing. The psychiatrist's personal analysis equips him to deal with his own needs.

Still another bridge is the psychoanalyst's concern with *problems of transference and resistance*. These phenomena are as certain to appear in social practice as they are in clinical practice. However, because of the greater complexities of dealing with the social system, they are not always as readily identifiable. The social psychiatrist brings to his relationship with an institution, a ready-made series of transference and countertransference problems. He may be seen as a magician or witch doctor with a capacity for working evil; he may be seen as a physician with a symbolic quality of healing. Most often, he is seen as the ambivalently regarded father. Thus, early in his interactions, he is either cursed or overloved. The group he deals with may turn to him with unrealistic expectations, and at the same time, secretly wish that he fail to meet them. When he does fail, he is castigated, and even excluded from further group transactions. This aspect of the transference relationship in the social interaction develops faster, is much less explicit, and is less dependent for its resolution on analysis of the transference fantasy than on the social psychiatrist's effectiveness in developing a perception of himself as a purposefully functioning human consultant.

Certain aspects of the countertransference problem must be stressed. The psychiatrist working in a social setting is even more likely to be involved in negative countertransference feelings than he is when dealing with individual patients. It is more difficult to be realistic about his position in the institution or community than his relationship to a patient. A distorted perception of the relationship can alter the dimensions of the contribution he has to make, and it may alienate his colleagues. Thus, his usefulness as a consultant will be vitiated. The psychoanalyst, by virtue of his training and particularly his personal analysis, should be better prepared to guard himself against these pitfalls than the nonanalyst.

Resistance to the psychiatrist in the role of consultant to the social system is probable. This is not the clinical role that, despite many misconceptions, the public has come to expect and accept. In the consultant role, the psychiatrist finds himself in muddier waters. He is expected to contribute his professional skill and also to behave as if he were an active part of the organization, with all the implications of involvement. As in clinical practice, there will be repeated attempts to force him into symbolic roles. The psychoanalyst, because he has dealt with these attempts in depth in individual therapy is likely to have developed the necessary patience and constraint with which they must be met in social practice.

The problem of resistance may be looked at from another, and perhaps more pragmatic perspective. No psychoanalyst needs to be told how stubbornly patients cling to old and damaging psychological patterns, nor why they do so. But it should be noted that change in social systems is resisted with equal tenacity. Institutions, for example, are often perpetuated long after they become outmoded and inadequate because they serve certain psychological, financial, social and other needs of persons associated with their administration and operation rather than the needs of those for whose service these institutions were originally designed. Those who have jobs at stake are loath to surrender the gratifications thus provided, regardless of the needs of the clients who might gain from institutional change; furthermore, they are usually unwilling to learn the truth about their resistance to change. The technical ability to overcome resistance in individuals may be the important catalyst for institutional change in response to public needs. Thus, what to some may appear to be a tenuous effect of psychiatry on personal or group resistance, may lead to permanent institutional change. The psychoanalyst recognizes the process of resistance more readily and is better equipped to deal with it patiently and with the knowledge that change, to be effective, often has to proceed very slowly indeed.

Perhaps the most significant of all these bridges is the psychoanalyst's understanding of the *dynamic transactional process* itself. There is a great similarity between the process of "working through of unconscious material" in the interaction between the psychoanalyst and the individual patient, and the process by which a group works through some of the unconscious resistance to the psychiatrist in the social setting. The latter requires the same kind of careful understanding as the former. The cues to the unconscious process which come from the individual differ from those which come from the group, but the fact remains that there *are* such signals. Psychoanalytic techniques offer guides for dealing with them.

With healthy people, one encounters a considerable amount of ego strength with which the psychiatrist can ally himself in order to expedite the working-through process. This corresponds, in psychoanalytic terms, to the concept of the therapeutic split of the ego. Awareness of, and competence with, the technique of modifying the ego permits the psychoanalyst to shift defense mechanisms and turn them to constructive advantage, with groups as well as individuals.

The psychoanalyst understands the importance of *development of rapport* with the patient in order to permit interventions to take place. Such rapport is also a necesssity when a psychiatrist consults for an agency or institution. Without rapport, no matter how knowledgeable he may be about mental health and illness, he cannot put that knowledge across in such a way as to make a genuine contribution to the group's efforts. Psychiatrists who lack understanding of what is happening between them and the significant persons in an organization frequently become overwhelmed with anxiety and have to withdraw from the situation.

It must be emphasized, in this connection, that the seduction of the psychoanalyst by one part of the personality, so familiar to analysts, has its counterpart in social psychiatry. Individuals or groups representing part of the social system will seek to seduce the psychiatrist in order to use him for their own ends, and to exclude him from working toward the overall ends of the total social system. Accordingly, the social psychiatrist must be clear about his role as consultant for the entire institution, rather than to a discrete part of it. He must remain a friendly but nonaligned consultant, so that he can intervene as necessary in any part of the system. In other words, the social psychiatrist must have the same kind of professional neutrality in dealing with a group or agency in the social setting as he maintains in dealing with an individual in the therapeutic situation.

The Person of the Analyst

Thus far, we have spoken of the contributions of psychoanalysis to social psychiatry largely in terms of theory and techniques. But the *person of the psychoanalyst himself* is a consideration that cuts across all others. A selection factor is at work here which must not be overlooked. *Among psychiatrists* those who seek psychoanalytic training are not content with simple causality but insist on probing the complexities. They have an affinity for piecing together seemingly unrelated facts and a sensitivity for the underlying meaning of behavior. Psychoanalysis as a profession both selects and trains the type of person who has the basic requirements for social practice.

The *personal analysis* of the psychoanalyst was already mentioned earlier as a factor of great importance. The personal analysis, as much as any other part of his training, gives the analyst his understanding of the dynamic transactional process. It has developed the self-understanding which makes it possible for him to be relatively honest with himself in the most delicate situations. He can, to a greater extent, withstand hostility because he understands and can handle his own reactions to it. He can withstand seduction because not only does he recognize attempts to seduce him, but he recognizes and understands that part of himself which is most vulnerable to seduction.

Any discussion of the psychoanalyst's own person must take note of certain relevant dangers. When an analyst deals with an individual patient, he concerns himself with his own character structures and defenses in order to achieve the objective neutrality so necessary to successful intervention. But when he deals with problems of a social system, there is a different and greater personal involvement, a personal stake in the social system, and presumably a commitment to it, all of which make it harder to maintain objectivity. He must ever be on guard lest in his absorption with the cause he serves as a social psychiatrist, he forgets the need for a continuing self-evaluation.

The social psychiatrist is more exposed than the private therapist. His personal vulnerability is greater because mistakes may be evident to a wide audience. Moreover, there are far-reaching social consequences of errors in judgment. The psychoanalyst's personal analysis should help him cope with the risks and the vulnerabilities.

Problems of the New Role

Like other psychiatrists, the psychoanalyst working in a traditional clinical milieu, has little need to concern himself with role definition. His patients' expectations of him, even when distorted by misconception, are based at least in part on their perception of him as a physician. This eases his task when he attempts to clarify his roles and functions for them. Thanks to long tradition, his students and medical colleagues are familiar with his role. But in social practice, no long-standing tradition offers role delineation and supports parallel to the traditional clinical milieu. The social field does not present him with patients in the traditional sense. Hence, it is difficult to perceive him there as a healing physician although his title may permit some of this aura to follow him into the social setting. Because it is a relatively new field, social psychiatry has little tradition; thus he must accept skepticism more often than deference, both from his own and from related professions as well as from the general public. He must know how he is received if he is to deal with the fantasies that may exist in the group with which he is working.

> Two Peace Corps examples of psychiatrists being introduced to a group of trainees offer different cues to the set of the group and its leaders toward the role of the psychiatrist. In one case, the introduction was "Here is our headshrinker." In the other "This is Dr. ———————, a psychiatrist who has been involved in planning the Peace Corps from the very first days."

The psychoanalyst's training will enable him to receive and to interpret with reasonable accuracy many far less obvious messages about himself which come from the group he works with. He must also take positive steps to contradict the negative stereotypes and unrealistic expectations which the messages bring him. He will have to build mutually trusting and respecting relationships with individuals in the group, in order to do his work and be of value. In time, and with infinite patience, he can achieve an adequate working definition of role, but until he does, he may feel as uncomfortable as anyone else who is suddenly shorn of a role identity which he has always taken more or less for granted.

In terms of intervention in the social system, the psychoanalyst must retrain himself to deal primarily with conscious data. Analysts commonly look for unconscious meaning; certainly, this is a necessity if one is to understand the dynamic transactional process. However, in social psychiatric practice, successful interventions utilize conscious mechanisms.

Psychoanalysts are accustomed to an attitude of calm and patient waiting. Social psychiatry, in contrast, often requires direct action. Furthermore, even within the limitations of the social psychiatrist's job itself, there may be a good deal of shifting from inactivity to activity and back again. Activity means more than verbal interpretation. For example, sometimes direct action may result in removing a disturbing child from a classroom. The university psychiatrist may intervene to help alter a student's course or timing of study, or field of concentration, rather than rely on therapy alone.

Activity may refer to the kind of situation in which the psychiatrist, sensing a group's unrest, may aid actively in uncovering its hostility in order to advance the group process. He then has to shift back into a sideline role of observation and evaluation, nonjudgmental, watchful, receptive waiting, able to maintain distance and impartiality. Otherwise he may find himself no longer a consultant but becomes instead another administrator.

All of this shifting about may cause the psychoanalyst some discomfort, particularly because he may wonder if it is valid for him to intervene actively or to shift his posture so frequently. In fact, the psychoanalyst shifts roles in his everyday life, although not necessarily within the professional context. He shifts from his role with patients to that with various groups of professional colleagues and in the institutions and organizations with which he is affiliated. He learns early in his professional life that he had better not play the "psychoanalyst" with his wife and children, with his professional colleagues, or with anybody else other than patients in treatment. The experience gained in these ordinary shifts of everyday life may make possible the shifts required by social psychiatric practice. Only a body of experience, and the development of supervision can substitute knowledge for intuition.

Supervision in Social Practice

We have observed throughout that there is currently no established body of training supervisors in social psychiatry. Yet the need for supervision in this field is as critical as it is in the clinical field and its lack is a tremendous obstacle to progress in the field—it affects the psychoanalyst as it does any other social psychiatrist. Without supervision, one can "bumble along" relying on analogies and intuition, and consult informally with other social psychiatrists. The classical *approach* to supervision—the detailed presentation of observed and felt clinical phenomena—needs no change for its application to social psychiatry. However, the *content* must be vastly different, and supervision will concern itself with different kinds of skills and knowledge, with awareness, for example, of social institutions, social systems, power structures, communication networks, and modes (and alternate modes) of entry into and intervention in the social system.

Supervision is needed to help the social psychiatrist to distinguish between individual crises and crises occurring in social systems. Moreover, it must help him to face the problems of entry into, and termination of, contact with the system. Here the problems are quite different from those of individual therapy, since the choice of intervention in social practice is different from that in individual psychoanalysis. Patients come to the psychoanalyst respecting the doctor-patient contract. In social practice no such contract exists. Termination of individual therapy occurs when an individual has grown in ability to cope alone with his problems. The goals of social practice may not necessarily be personal insight, but rather the initiation of a social process which will result in positive action. With an agency or community, termination of a single intervention occurs when the organization is capable of dealing with the problem at hand without the presence of the psychiatrist. However, consultation may well be an ongoing process and the psychiatrist is "on call" or "on retainer" and never in fact terminates.

Psychoanalytic training and experience alone do not create sufficient skills to deal with social psychiatric problems. The analyst must build onto his skills by adding approaches and information from many disciplines, which are less relevant to his primary tasks as a psychoanalyst. For example, the psychoanalyst in traditional practice rarely has the opportunity, or motivation, to observe and study the problems of socioeconomic groups lower than the middle classes, and certainly not those of the underprivileged minority groups in particular. Once he is to interact with *total* social systems, his effectiveness will be severely restricted by lack of knowledge of these groups, and their social and physical settings. In social practice, the psychoanalyst needs further training and actual supervision from people other than psychiatrists. Supervision may have to be a shared experience between supervising social psychiatrists and sociologists, community organization workers, political scientists, economists, lawyers and many others. Until there is far more certainty in this field, supervision may have to be ongoing, not just part of the training process. Continuing evaluation of problems ongoing sharing of experience will be necessary to create a useful body of knowledge.

Summary and Comment

Among psychiatrists, the effective performance of psychoanalysis in social practice is attributable to broad and highly specific relationships between the theory and technique of psychoanalysis, the concepts and practice of social psychiatry, and to considerations referable to the analyst's own person.

While we have undertaken to demonstrate that there are many bridges between psychoanalysis and the social practice of psychiatry, we would be loath to claim for psychoanalysis alone the preserve of social psychiatry. Indeed, there are many practitioners who make extremely important contributions to this field, some without a medical or psychiatric background. They can do so because of their broad experience in the community in understanding social dynamics as well as the processes of change.

In fact, many psychoanalysts are neither cut out for social psychiatry, nor interested in different kinds of activities and concerns that make up its practice. They will prefer for a variety of reasons to practice in more traditional therapeutic fields. The future of social psychiatry is thus equally the broad domain of the concerned psychoanalyst and psychiatrist, as well as the social scientist. For it appears that the development of the field will come through joint endeavours, in which the identities of the various disciplines will blur, and from which will arise a new kind of practitioner and researcher.

Clearly, the ability to assume responsibility "for the mental health of the population of the United States" will require changes in professional orientation and practice. To meet these new demands, the psychoanalyst will be supported by his professional knowledge and training; by his ability to be flexible at the way he sees and tackles problems; and by the techniques he has at hand to modify established thinking. Finally, the challenge of this presently uncharted area of professional practice exists not merely in the ability to make voyages of discovery, but also in the knowledge that the fruits of this New World will, with certainty, enrich the old; that current individual psychiatric practice will be enlarged by the knowledge and skills gained in the field of social practice.

REFERENCES

1. Leopold, R. L., Duhl, L. J. and English, J. T.: A Mental Health Program for the Peace Corps. Human Organization, Vol. 23, No. 2, Summer 1964.
2. Stanton, A. H. and Schwartz, M. S.: The Mental Hospital, New York, Basic Books, Inc., 1954.
3. Caudill, W. A.: The Psychiatric Hospital as a Small Society, Cambridge, Harvard University Press, 1958.
4. Fried, M.: Social Problems & Psychopathology, in Urban America and the Planning of Mental Health Services, Symposium #10, Group for the Advancement of Psychiatry, New York, 1964, p. 436.

PSYCHOANALYSIS AS INTENT: IMPLICATIONS FOR COMMUNITY MENTAL HEALTH

By ALAN B. TULIPAN, M.D.

PSYCHOANALYTIC INSTITUTES have generally tended to relegate community psychiatry to the position of a distant relative, entitled to its own existence and useful in its fashion, but without major moment in its effect on psychoanalysis nor, conversely, able to gain from the sanctified insights of psychoanalysis. The aim of this presentation will be to offer a conception of psychoanalysis which can allow for a broadening of its usefulness into other interpersonal arenas, particularly through community mental health practice.

Community psychiatry is not alien territory for the analyst. Analysts have frequently conjectured means whereby their insights could save the physical and interpersonal world from its relentless march toward doom. The literature abounds in conceptions concerning the community, some worthy and well-wrought, some eminently impractical but all with manifold social and political implications.

Difficulties have arisen largely because that which has seemed sound theoretically has had no practical possibility of success, or because the ideas promulgated were grandiose push-button solutions to a cosmic or cultural problem which would defy such ministrations. The idea of affording psychoanalysis to world leaders, especially to those with their finger on The Button, is intriguing, but I have some question as to whether Mao Tse-Tung would be an accessible or well-motivated patient.

Community psychiatry is an attempt to bring to fruition changes in individuals through personal contact or through indirect intervention by way of the environment. Many analysts balk at the idea of departure from their usual modality of treatment and functioning, and this is understandable. The economic and emotional investment in the training process, the tendency for us all to remain embedded in our way, the interactions and gratification accruing from our relationship with patients and the professional stimulation gained tend to discourage major conceptual or practical departures.

Psychoanalytic theories, in general, have been directed to the personal or interpersonal metamorphoses effected in the one-to-one doctor-patient situation. I would suggest, however, at the risk of becoming cosmic myself, that if his theoretical groundwork is sound in its conception of interpersonal processes, the analyst can expand theory to include a unified orientation to other interactions, including those of large groups. Further, interpersonal growth in the psychoanalytic relationship is based on dynamics that can be found in modified form in the larger arena of the community. Conversely, the enabling modalities used in community psychiatry have rough counterparts in the individual psychoanalytic process. If it is alien for the psychoanalyst to enter into these, it is alien for him to involve himself in a large part of what he does in the office.

The analyst, probably more often than he knows, does enter into the life of his analytic patient by direction and indirection. The judgment implicit in a raised eyebrow, a voice inflection, or even in a pregnant silence can engender a direct response of action by the patient which may ultimately be taken by both patient and therapist as the product of a remarkable insight. In recent years, this intrusiveness has more and more been uncovered experimentally. Such subliminal directiveness on the part of the analyst we need neither condone nor disparage. It is there, and will probably continue to be there.

But there are other times when direct intervention is a conscious process; when preventive ministrations (education about contraception in a teenager), emergency services (for psychogenic urinary retention or debilitating anorexia), referral to other psychiatric services (hospitals, etc.) and home visits (with parents, spouse, etc.) are all part of the practice of even the most strenuous advocate of the one-to-one relationship.

Yet, often these activities are seen as adjuncts to psychoanalysis rather than as a reasonable part of the analytic process; as corollary functions rather than as an interaction directly tied in with the analytic relationship. I believe, however, that analysis does not stop at these times, except for those who see analysis as a technical feat in which the possibilities of the broadening of interaction are delimited by preconceived boundaries. I would like to present a view that points up the fallacy of looking upon psychoanalysis as a technique with any preconceived methodology and that opens up at the same time some horizons wherein community psychiatry becomes as reasonable a part of the armamentarium of the psychoanalyst as the couch and the 50-minute hour.

First, let me indicate what I think psychoanalysis is not. It is not some *thing* to be administered as an across-the-counter commodity, nor is it a technique applicable to some patients with certain types of problems. I am not even certain that it requires the act of bringing concepts from unconsciousness to consciousness, though this is often incorporated into the definition of the term. Every analyst, for example, has witnessed a profound and seemingly permanent change in a patient without knowing why this has occurred. In this connection, I have heard it said that it is far easier to account for failures (in psychoanalysis) than for successes.

In short, I believe analysis should be considered not as a technique, but viewed rather in terms of its interpersonal intent, an intent that might well prevail in other good, growth-producing relationships, but with its own special possibilities and limitations. The fact is that all interpersonal activity has special built-in possibilities and limitations, dictated by the personal attributes and qualities of the participants. The result, of course, is that each relationship is unlike any other.

However, all psychoanalytic relationships, and, in fact *any* relationship (parent-child, marital, teacher-pupil, etc.) has three intermingling potential phases: (1) an interpersonal encounter, with all its conscious and unconscious ramifications, (2) reorientation to self and world, and (3) activation toward change in the physical world. These three aspects of the relationship bear a dynamic, but not necessarily a fixed temporal relationship to one another. Changes in any one can set off changes in the other. They exist as phases of all interactions. The more any relationship approaches an intent of growth, healthy encounter, and rationality, the more it resembles the psychoanalytic encounter.

The greater the possibilities for positive experiences of expression and interchange in the encounter, and the greater the possibilities for continuing growth through the reorienting process, and the more the potential for activating oneself as a result of internal reorientation, then the greater the chance for the fulfillment of selfhood. All of these three phases are positively experienced in the relationships of healthy, self-affirming people.

The psychoanalytic process is a telescoping of such interpersonal experience in which there is generally a greater *consciousness* of change as its intent, though the modalities of encounter still fall within the framework of the encounter-reorientation-activation triad.

It is in the psychoanalytic encounter and in the reorienting process that differences of opinion have arisen among psychoanalytic practitioners as to methodology. I would contend that in the frame of reference presented here, there can be no methodology, since this would hinder rather than help the potentially self-affirming process. The imposition of technique would: (1) hinder the truth and the scope of the encounter, (2) limit the potential for the patient's reorientation toward self and world, (3) impose strictures on his possibilities for self-activation.

A rich and meaningful personal *encounter* in psychoanalysis requires, aside from physical presence (and its concomitants, speech, gestures, and expression), other potentially binding interpersonal forces, such as (1) relatedness without automation and without preconceptions, (2) exploration for the sake of knowing and understanding, (3) the willingness to withstand the anxiety that accompanies change or the vision of its possibility, (4) confrontation of an honest, nonevasive kind, and (5) the willingness to help and to enable, where this is not a condescending maneuver.

For such an encounter, at least one person should have the special qualifications that might make this possible. In the psychoanalytic situation this is very often the analyst. He must be relatively free to maintain an integrity and dedication devoid of prestige and power motives. He must be willing and strong enough to withstand the anxiety and even the despair that accompany changing attitudes, orientation and failures. He must rid himself of exploitative modes of encounter or at least know of their presence when they occur. He must be capable of a constant relatedness to what is going on in and outside of himself. He must be free enough to be honest, nonevasive, and open to new and unique aspects of encounter.

The second major phase of psychoanalytic treatment is *reorientation*, wherein the patient, and, in part, the analyst, gains a new image of himself which is closer to reality, to potential, to self-affirmation. This does not imply a change toward arrogance and self-righteousness or a new role-playing as is so frequently seen as the result of some therapies. Reorientation can be achieved in two ways: through the intellect and through the encounter itself. The instruments of analysis are generally verbal and intellectual, just as the instrument of child therapy is play. Often the instrument is misunderstood to be the reason for the change rather than the means toward a therapeutic encounter. A patient can gain some degree of reorientation from the verbalization and repetition of the nature of his patterns of living in the present, with corroborative evidence from knowledge of past patterns and their origins. This is, at the least, one way to learn what to look for in one's interpersonal activity, and there may well be flashes of recognition signaling a reorientation after repetitive intellectual corralling of the facts. In any case, it would be economically wise for us to maintain this view since I doubt that patients would respond to our fee requirements were we simply to play checkers, or, with some of our intellectually better endowed patients, chess. Batting balloons in the playroom is infinitely more fun and may bring to light patterns quite similar to those obtained by talking, but it does reduce the expected dignity of the professional encounter.

I do feel, however, that the very encounter itself is often far more productive of significant reorientation than the words. It offers a new, intense, intimate and important relationship with new possibilities for the patient in his drive toward mental health. He here meets with responses of a kind which he either did not receive, or received but misinterpreted in the past. He is now in the situation where, hopefully, the persistence of noncondescending attitudes by the analyst will ultimately help him revise the patient's distorted self-image. Through the persistence of new reflected appraisals, he becomes reoriented to his interpersonal possibilities, and the reorientation serves as a bridge to new orientations in other encounters.

However, reorientation is not by itself enough. The patient must take the third step—*self-activation.* He cannot reorient himself in a vacuum and gain. He must transcend his past in the physical world, step away from his previous patterns, acquire some immediate or long-term commitments in his life and work, stemming from an inner motivation. Last and most important of all is his willingness to withstand anxiety for the sake of growth, and new encounters, and ever widening orientations.

This anxiety is a link between the three phases of the relationship. Without it there is little likelihood that reorientation or activation or change will take place at all. Anxiety is the normal accompaniment to confrontation with the new and different or the vision of its possible presence. The healthy person will transcend his anxiety and move into the continuum of the future despite it. The unhealthy person will dig his groove deeper through the use of any of the devices that are, in fact, his illness. Anxiety is a human experience and in itself need not be inhibiting. Inhibition occurs when the patient lacks the courage to transcend his anxiety. The courage for transcending marks the mature man as retreat is the mark of the immature. In psychoanalysis there are always new confrontations, and, therefore, a constant struggle with anxiety, as is true in any mutually self-affirming relationship.

Psychoanalysis has been looked upon as a treatment process in which the encounter involves the uncovering of rational and irrational attitudes to the analytic encounter itself as well as to other encounters past and present. The reorienting process is primary.

Traditionally, non-psychoanalytic psychotherapy would place less emphasis on the uncovering aspects of the encounter, more emphasis on activation, either self-imposed or externally imposed. Thus, the enabling is of greater moment, and this, in turn, may hopefully lead to some changing orientation to self. Enabling psychotherapy would be primarily activating. Reorientation is its secondary goal.

The psychoanalytic process is never set. At the beginnings of psychoanalysis, the encounter was delimited by scientific design. Since participant-observation has become a fact of psychoanalytic life, obviously the encounter has become more active and necessarily infinitely variable, since no two encounters involving real participation can be the same. There can be fewer and fewer preconceptions, since participation implies spontaneity and necessarily the presence of "surprise." The more the preconceptions, the more deceptive and delimited the encounter. The truth of an encounter is likely to be analyzable through postconception rather than through preconception.

That is, the accident, the surprise happening, the uniqueness of responses within the encounter become the significant material for a two-person appraisal of the patient. The nature of the encounter is probed after the fact rather than before. Otherwise, that aspect of the patient which is unique might well be missed. The health of the analyst, his selfhood, spontaneity and willingness to examine serve as the base upon which the patient's mode of encounter can be appraised and also against which the patient can test new modes of encounter through self-activation.

Thus, preconceptions interfere with the manifold "truths" of the specific two-person relationship, and particularly, of the patient's general ways of relating. What is set up by the analyst all too often is a relationship between the patient and the analyst's preconceived ideas, rather than between the patient and the analyst. Fortunately, very often the sensitivity of the patient and his own positive drive toward mental health lead him to see beyond this and to relate to the person of the analyst despite the latter's attempts, conscious or otherwise, to befog the scene.

Reappraisal of self and reorientation to one's position in the world certainly can take place without conscious awareness at all, as is obvious in the normal growth and development of children. In fact, it is very often questionable whether what is *thought* to be uncovered in the analytic process really makes for the ultimate change. It is more likely the *intent* of rationality and honesty and the special qualifications of the analyst to maintain these qualities interpersonally that makes for the change in patients, rather than any aspect of technique or preconception. The analyst does not, indeed cannot, restrict his role to the simple uncovering of things taking place in the present or the past. He comes through, at times, to be sure, as an uncoverer; at others as a questioner, challenger, responder, reactor, but also, as a helper, enabler. Of course, with any given patient, one or another of these characteristics will predominate. But this is because encounters *are* different (even in psychoanalysis), not because they are designed to be different.

In each encounter there are new "possibilities" and it is those possibilities which make for the large share of difference among relationships. The possibilities for individual unfolding are different when I am involved with a 3-year-old child, a companion at a cocktail party, a group at a convention, or a patient in psychoanalysis. This does not imply that I am any the less myself in each of these circumstances, but rather that my responses are modified by the nature of the encounter and the nature of the surroundings. My individual freedom and selfhood are in no sense jeopardized or changed. I am myself relating to different people in different situations in differing ways because "possibilities" of becoming and unfolding are changed or delimited or broadened by these personal and situational factors.

I am similarly changed, delimited and broadened by the "possibilities" offered by different patients in psychoanalysis. This is not a matter of a different "approach" to a patient which implies a preconceived kind of role-playing or mathematical formulation. Were we to see a patient who had the tendency to "approach" people in his everyday life through a calculated assessment of the gains that would accrue from the approach, we would label him a compulsive, unspontaneous, alienated man. We cannot afford the freedom to be similarly pathologically encumbered when we perform our professional function. We don't "approach" our patients in specific ways. We are ourselves and are relating according to the quality and possibilities of the relationship. Thus we do not delimit. We are what we are . . . including all possibilities for people to be. Limits are imposed by the social scene, not by ourselves in our relatedness. We are limited only in relation to our internal possibilities as modified by past experience and present personality. This is played against the external world as another modifying force. Since we relate as people, not as automatons, we have many things to offer patients.

Because of his own experience, limitations, previous expansion or sophistication, a patient might need greater or lesser guidance as such, from the analyst. He may, because of the severity of immediate discomfort, require medication or referral to another medical specialist or hospitalization. Some will have to be given direct advice. This is not alien to the role of the psychoanalyst. The analyst is a helper when help is rational and when it does not interfere with his capacity to assist the patient to achieve interpersonal rationality.

A patient who is acutely suicidal cannot be met by the impassive mien. He needs hospitalization or drugs. A mother who cannot care for her family needs practical aid. Chemotherapy need not be spurned by the psychoanalyst as an adjunct to his patient's treatment. He can also take the role of a rational authority even though it might, indeed, suggest that the patient is inexperienced in certain areas (which he is), or though it might engender an old reaction pattern. Even when this happens, it can be used to point up the fact and nature of the pattern.

Some patients who have led an insular life have a right to "know" just purely on an educative level that things are conceivably different from that which their own experience has revealed to them, and perhaps to know it sooner rather than later.

In many situations of crisis, the analyst who sits back impassively because of his adherence to a technique is either obsessively subservient to a textbook approach to analysis or is excessively inhuman. The fact is that the therapist offers help, if help is needed, whether he be psychoanalyst or otherwise. This is true if only because he is a human being with a helping orientation. If he happened to have been caught in an interpersonal trap, so be it. Let him be free enough to be able to recognize the possibility and to bring it out.

The arena of psychoanalysis is broadened in proportion to the freedom in the psychoanalytic relationship for interpersonal patterns to unfold. Thus, an insular relationship allows for a meager display of interpersonal patterns and therefore less is uncovered. A rich relationship, conversely, has greater potential for the uncovering process, with vast opportunities for corroboration.

If psychoanalysis is not so much a method as it is a conception; if it is a creative way of orienting oneself to an interpersonal situation in order to penetrate to its ultimate truths, then the analyst does not become any less an analyst when he sees a patient once a week than when he sees a patient three to five times a week. His relationship to a patient is psychoanalytic primarily through his capacity and his conscious or unconscious intent to be free enough to make a clean contact with another human being, uncluttered by interpersonal deceptions. He desires to make this contact as meaningful as possible, within the built-in limitations at any one time. The fact that an encounter with a patient occurs once a week, as in a community psychiatric clinic, rather than three times, in itself may change the therapeutic situation, but not because of a difference in technique or intent, or in the selfhood of the therapist. Under that circumstance, it might well make interpersonal sense that the encounter be more enabling, less uncovering. Each patient has unique things to offer and unique ways of receiving in a therapeutic encounter. He may be capable of abstracting or he may be concrete. He may need direction or he may not. It is condescending to, and disrespectful of a patient to label him as "non-psychoanalytic" as is often done with so-called "blue collar" people, or those who come to community health centers. False expectations (either too high or too low) interfere with the molding of any potentially productive, self-affirming relationship, whatever the technique or the frequency.

The psychoanalyst is most helpful, be it to a private patient, clinic patient, relative, or friend, if he encounters that person with the intent of rationality and interpersonal honesty. Further, the psychoanalyst will be especially helpful personally if he has the capability of reducing his role-playing to a minimum, is free to make new, unique and honest encounters, has the will to transcend rather than retreat before his anxiety, and aims for relatedness with others neither with attitudes of condescension nor with expectations beyond the possibility of achievement. His professional discipline arises from his commitment and courage to maintain this difficult stance with the unusual intensity necessary in his every professional contact.

From these considerations it is a relatively short leap to the question of how the psychoanalyst relates himself to the community and to community mental health. A community or its individuals can be met with a therapeutic intent that does not differ from the intent that prevails in private psychoanalytic practice. The built-in limitations are there, the creative mode of encounter pertains, the goals are, in truth, no different. Differences are solely in the realm of the realities of the personal and group resistances, the possibilities and the situational limitations.

Community mental health has many operational arenas. Its scope is wide, its possibilities limitless; it comprises, in general, the following:

1. Therapy within the four walls of a clinic; individual psychotherapy for adults and children, group psychotherapy, chemotherapy, casework.
   Because of the general prevailing patient loads, it is unusual for individual treatment to be more frequent than once a week. The patients are often of a socioeconomic group different from that of private psychiatric patients. They are more symptom-oriented, less sophisticated on the whole, more likely to seek immediate comfort as they enter treatment, Interactional possibilities are often modified as a result. As hospital patients are discharged on pharmacotherapeutic agents, more outpatients are in the diagnostic category of the psychoses.
2. Expansion of the community clinic into other modalities for working directly with individual patients: Day-care centers for disturbed children and adults, broadened resources for the aging, night hospitals, twenty-four-hour emergency services, rehabilitative services, consultative facilities.
3. Ancillary services of teaching and/or guidance and consultation with key professionals and semiprofessionals in the community who see emotionally disturbed patients on a direct and regular basis, such as nonpsychiatric physicians, the clergy, public health nurses, school personnel and others.
4. Wider community activities designed to capture, in a progresssive, self-affirming program, groups of people who cannot be captured by the usual psychotherapeutic means or by the corollary means indicated above. This would involve, for example, the implementation of programs to "meet" youth on a level that is meaningful to them; to initiate rational activities to deal with delinquency; to develop, together with other key individuals and groups in the

community, coordinative, nonduplicating efforts; to "spot" potential emotional problems before they burgeon into difficult overt ones; to enhance the "possibilities" for the individual by helping to enhance the community itself; to help minimize community problems in housing, segregation, and employment so that the mobility of its citizens is increased, and hope replaces hopelessness.

In effect, the people in the community, and the community itself, may be as unsophisticated, as resistive, as symptom-oriented, as interested in immediacy, as hopeless as many of our individual psychoanalytic patients. The means of coping with this are not vastly different from the means of coping with similar attitudes in individual patients. As one "meets" individual patients differently because of individual differences and possibilities, so does one "meet" the community and its people on the basis of its differences and possibilities. The principles and movements described which are of significance for analysis can be applied to community psychiatry. The three phases of (1) encounter, (2) reorientation, (3) self-activation are specifically pertinent. And the analyst can use his special qualifications to enhance their implementation and effectiveness. Members of the community, sometimes those in positions of influence, so often make unhealthy encounters with other individuals built on attitudes of condescension, immediacy, expediency, and irrational authoritarianism, that they engender a relentless hopelessness in the group or individual psyche. New encounters, if available, can often cause some blue sky to appear on the horizon and make for the possibility of attitudes of hope.

The professional person has the opportunity to make an encounter with the community himself and also to contribute to the opening up of areas in which healthy encounters can occur for its citizens who might otherwise be deprived of them. He has to meet the community in such a way that (as has been expressed in relation to the individual) he is neither condescending, nor irrationally overexpectant. If he is either of these he will be minimizing his effectiveness.

His own encounter with the community is enhanced by the elimination of preconceptions, so that he can meet it by way of its uniqueness, its possibilities, and its accessibility. As with individuals, each community has its own personality, its possibilities for expansion, its limitations. A climate of opinion, an insidious factionalism, an entrenched political hierarchy, a socioeconomic disparity, together with other subtle and unsubtle characteristics can limit the effectiveness in one community of an approach which might work in another. Projects have often failed because they are either ill-conceived, ill-timed, inappropriate, or even anachronistic. In short, those initiating the project have failed in their encounter with the community to accept its uniqueness. A good example of this would be the attempt to give the predelinquent or delinquent positive experiences in life by offering him some such activity as basketball as a Saturday night recreational activity. Most delinquents I have known would scoff and proceed posthaste toward what to them would be far more stimulating pursuits.

The enabling function appropriate to individual treatment can be expanded to include the community. There has come to be such a built-in sense of helplessness among the desperately deprived lower socioeconomic families within many of our urban communities, that the only goals in any way attainable are immediate ones. Optimism associated with a feeling that anything is worth waiting for is nonexistent, and life is built around the gratification of immediate desires or the attainment of quick pleasures. Within such an atmosphere, it is impossible to learn and grow, since learning and growing require an awareness of future possibilities. Nor is it possible to develop the potentiality for the patience and the compromise which are a part of a growing relationship since these, too, require the optimistic feeling that something better may be forthcoming. So children become burdens rather than people, parents become encumbered by feelings of futility and a cyclically oppressive climate pervades their being. It is this hopelessness that, for example, moves the delinquent to seek out his shiny immediate gains, to fend off our attempts to implore his entrance into psychotherapy, and to view us as suckers and squares when we exercise restraint and compromise. Obviously, we have failed miserably in capturing such individuals in treatment as we have defined treatment in the past. Yet, were we to consider psychoanalysis from the point of view of the processses of encounter, reorientation, and activation, would there not be broadened horizons for our usefulness in this area? Each element of this triad has an effect on the other, and often in dramatic and predictable ways.

Thus, political or social reform can so enhance the encounter between the individual and his community that profound changes can be effected in those previously unreachable. The most obvious example of this, of course, is the sense of hope injected into the minds and lives of so many people by the Supreme Court decisions of the past ten years and its subsequent impact on American life. The prospect of mobility, where no such prospect existed, the potential for self-activation, have changed the orientation of millions of people from despair to hope. The injection of similar attitudes, possibilities, and growth-producing relationships is possible within each community, with a resultant gain in the mental health of its citizens.

These new encounters, together with a flowering of community resources can provide the open space for new orientations to occur to self and to world. The more limited the "possibilities" in a community, the more limited the new appraisals and assessments. An insular world breeds personal insularity.

With the opening of new channels in the community there is, then, the chance for new modes of self-activation among its people. For example, if there is a need for productivity on the part of a teenager who comes to see its advantage for himself, he cannot activate himself until and unless there is vocational opportunity, apprenticeship training, job opportunities, organized employment bureaus, vocational testing, expanding paid-employment opportunities and an accepting orientation by local merchants, police, school authorities and other community groups. The dignity of the community very much affects the dignity of the individual who encounters it. The counterpart in individual treatment is seen when the new significant encounter leads to a reappraisal of self, and the interpersonal possibilities have suddenly burgeoned. Wherever the psychoanalyst can use his special skills to contribute to the opening of such new channels in the community, he is indeed operating within the framework of psychoanalysis toward the improvement of mental health.

To summarize: The method of the analyst, if there be method, can never in any two of his encounters be the same, because he must meet each patient on a level where encounter is possible, where reorientation can take place, and where the patient has room to move and the courage to move by self-activation. His special contribution arises from his ability to be open and flexible, honest and rational in each new encounter, and devoid of obfuscating preconceptions. His professional intent, therefore, becomes his most significant tool, and it is a tool with many uses. Individual psychoanalytic psychotherapy is one. The enhancement of the mental health of the community is another. But, this can become a part of his psychoanalytic practice and function rather than an alien activity only when he abandons the view that psychoanalysis is in any sense a technique to be administered as an across-the-counter commodity.

THEORETICAL AND PRACTICAL ISSUES IN A COMMUNITY PSYCHIATRIC PROGRAM

MONTAGUE ULLMAN, M.D., LESTER A. GELB, M.D., WILLIAM F. WEBER, M.D. AND MARK TARAIL, D.S.W.

The thoughts and ideas to be presented have been generated out of a three-year experience in the building of a community-oriented psychiatric program based in a general hospital. In the absence of an inpatient psychiatric unit, our initial efforts were directed toward the development of preventive, diagnostic and treatment services for three types of communities—the population which makes up our outpatient unit, the intrahospital population of medical and surgical patients and the community-at-large.

Our base of operations is Maimonides Hospital, a 560 bed general hospital situated in a lower-middle class section of Brooklyn. In the phasing of our program, initial efforts were devoted to the development of staff and the extension of outpatient services, the organization of a liaison program, and the creation of a foundation for the extension of services into the community.

A review of the activities of the first three years of operation has led to the identification of a number of practical problems that undoubtedly keep recurring in many different settings and a number of theoretical issues which have relevance to some of the new vistas emerging upon the scene of contemporary psychiatry.

The Outpatient Community

Here the main problem is an old one, but since it has never been adequately set to rest it presents itself anew to each one who begins to concern himself with the psychiatric needs of those patients who cannot afford to present themselves for private treatment. We are referring to the overwhelming demand for services—that dormant need that springs to life whenever a new psychiatric facility appears. Where does it come from? Why are so many people so disturbed?

The next obvious question, of course, is how to obtain the manpower to deal with the problem. Granted that the physician has a leadership role in the treatment of mental illness, why has it proven so difficult to orient the physician-psychiatrist to the public health aspects of his own specialty? The divorce between the public and private domain is nowhere as great in medicine as it is in psychiatry.

The third set of problems arises in connection with the need to revamp one's treatment style in order to develop flexible strategies that will be at least partly efficacious in the setting in which they are applied. Does one turn to limited, short-term contacts with patients with a sense of defeatism and in a spirit of compromise? Or can one, by a process of selection, employ this as an optimal form of therapy for certain patients? What kind of internal rearrangements does this necessitate for the psychoanalytically trained psychiatrist?

These and other questions had to be faced as we sought to devise treatment methods applicable to broader segments of the population and in larger numbers than those seen in private practice. In the clinic we are more concerned with the question of what life brings to the patient, whereas in private practice we have often limited ourselves to what the patient brings to his life situation. In psychoanalytic practice, we make ourselves available to the patient for the purpose of exploring and expanding his capacity for intimacy, honesty and freedom in a human relationship. In the clinic situation we are more apt to act as facilitating or catalyzing agents in improving his relations with the significant figures in his life, so that they, rather than the therapist, emerge as the chief recipient of the patient's initial experimental efforts at self-realization and growth.

The Intrahospital Community

Certain problems were immediately obvious as we addressed ourselves to the needs of the intrahospital community of patients, doctors, nurses and other ward personnel. The situation seemed wrong in a number of ways. With few exceptions, relations between psychiatrists and other physicians tended to be disjunctive rather than truly collaborative. The management of consultation requests tended to follow the model of responding to consultations as set down and developed in other medical specialties. No one seemed to take cognizance of the fact that, as Meyer and Mendelson[1] have shown, the difficulties extend in ever widening concentric circles about the patient to include the family as well as medical and paramedical personnel.

Young physicians are still not taught to think in interactional terms. They are aware of the ministering aspects of medical practice, not the transactional aspects.

The patients seen in the liaison program come from different social, cultural, educational and economic backgrounds than the psychiatrist and hospital medical staff. They are usually physically ill. They rarely consider their problems to be functional in origin, and if they are aware of emotional difficulties, they do not feel that they bear on the current problem. In short, they are rarely motivated for treatment in the psychoanalytic sense of the word. The psychiatrist is deprived of his usual professional accoutrements. Privacy is often impossible. Interruptions are frequent and time is short. The patient's physical condition and its treatment have priority. As for the hospital staff (physicians, nurses, aides, etc.) their focus is organic, with functional problems often seen as an inconvenient intrusion. Their own interests, background, and experience may make understanding of, and communication with, the patient difficult. There is often lack of awareness that what is perfectly obvious to them may be unintelligible and frightening to the patient.

In response to these considerations our liaison program has been undergoing a gradual, but thorough, restructuring, aimed first at greater care in the identification of the pathological focus and second at providing the physician with a better base from which to attempt to cope with the behavioral complications of the patients under his care. The investigatory team has been broadened to include the medical social worker assigned to the ward and a psychiatric nurse assigned to the liaison program. The former would have available background social data and an assessment of the current life situation whereas the latter would be knowledgeable about the patient's behavior from the point of view of the nurses and other ward personnel. Every consultation is reviewed by this team in order to identify the contributory factors in the situation, to formulate an approach to its resolution and, most important, to work with ward personnel at all levels toward a resolution. It was only in a relatively small number of instances that the responsibility for the direct care of the patient had to be shifted to the psychiatrist.

The Community-at-Large

So far, what we have been talking about is psychiatry in a general hospital setting. The qualitative jump into "community psychiatry" occurs when we look beyond this and begin to engage directly with the community and in turn apply what we learn to the intrahospital community. We moved into this terra incognito armed with three sets of assumptions. The first was that the problem of mental illness was so pervasive that the struggle against it could be successful only to the extent that the professionals in the field could evoke the support of the community at a grass roots level. Second, our orientation involved a reaching-out approach both in terms of education and prevention and in terms of exploring the social settings in which patients worked and lived. The final tenet follows from this, namely, that social therapy, defined as planned intervention in the social milieu of the patient for the purpose of ameliorating behavioral patterns in the direction of healthier modes of living, has to articulate with the practice of psychotherapy.

When we addressed ourselves to the community-at-large, we found what, unfortunately, we have all come to expect to find—an anarchic, chaotic patchwork of services. In sum, these amounted to a desperate but inadequate holding action, this despite isolated patches of heroic enterprise.

As community psychiatry has developed in our Department, the term is used to include two sets of operations. These comprise our growing program of extramural community mental health activities in the areas of education, prevention and rehabilitation, as well as the structuring of diagnostic and treatment services in accord with our community goals. Our approach extends out to the life of the patient in the community. Members of the treatment team expand the areas of their operation to include problems existing in the family milieu, vocational guidance, job training and finding, the restructuring of recreational and social activities of the patient and community planning.

A few examples from our experience at Maimonides Hospital may serve to illustrate such reaching-out activities.

1. Responding to the needs of two groups in the community who sought our help, we have initiated a program involving the inservice mental health training of guidance counsellors and attendance teachers in public schools in Brooklyn with a view to increasing their skills in coping with disturbed children and their families in the school setting.

2. We have instituted a training program to explore and demonstrate the role and function of the psychiatric nurse in a general hospital-based community psychiatric program.

3. In Brooklyn, as in most other communities, there are a number of public and voluntary agencies in the mental health field. There are no channels at present for coordination, cooperation or joint community planning among these agencies. We are currently exploring ways of developing informal as well as formal channels for communication and planning among a variety of different mental health and related agencies in a given locality.

4. In order to develop professional personnel with skills in community planning, we have established a unit in the Department of Psychiatry, in collaboration with the Columbia Schools of Social Work, for the training of social workers as community organizers in the mental health field.

Theoretical Issues

The problems we have touched upon are part of the ferment which pervades all phases of our activity, not sparing traditional approaches to treatment, training, and organization of services. Yochelson[2] describes the situation very well in his discussion of sectarianism in psychiatry:

> "The growth of community public health and social psychiatry momentarily has disturbed the homeostasis of general psychiatry in such a way that as a listener at some national meetings on this subject, I was left with the distinct impression that a number of colleagues were certain of this: social psychiatry is antithetic to the treatment of the individual. Indeed, some departments of psychiatry and mental health groups have experienced conflict over the apprehension that social psychiatry pays no attention to the individual patient. This makes as much sense as criticizing a physician for his efforts at locating the course of typhoid infestation instead of restricting his efforts to the treatment of the individual typhoid patient. Both social and individual approaches are essential in the interests of treating the total situation. However, when the sectarian attitude prevails—that is, when each proponent of his own interest feels defensive to the extent of downgrading or eliminating the other—we pay a sad price. The validity of various psychotherapeutic techniques, particularly that of individual psychoanalysis, is not at all in question. Rather, from the point of view of total good, a public health oriented psychiatry must supplement (not supplant) individual psychiatry. Should the sectarian attitude of either orientation (community or individual) prevail to the exclusion of the other, the profession of psychiatry never will attain full maturity."

Lebensohn[3] identifies the problems with regard to training:

> "The fact must be squarely faced that the present-day undergraduate and graduate training tends to direct the young psychiatrist into the more remunerative, less arduous, and more prestigeful pastures of private practice

and away from the rougher challenges of research and caring for seriously sick patients. The hard truth is that we are expending enormous effort, time and money to train a highly specialized group of experts who will devote the bulk of their professional careers to a small group of highly selected private patients. It is a disturbing truth when we view it in relation to making more than a dent in the core problem of severe mental illness in America."

Rosenbaum and Zwerling,[4] in a frank discussion of the impact of social psychiatry upon a psychoanalytically oriented department of psychiatry, attempt to place the problem in its proper perspective by emphasizing that the difference between individual psychotherapy and social psychiatry is simply a difference and does not harbor oppositional features. They pose as the issue, not the relevance of one to the other, but the manner of their articulation. After analyzing the various resistances that have arisen within their own situation they end with a plea for a greater theoretical interplay.

It is in the interest of seeing this kind of interplay come about that we wish to examine briefly the problem of how the two areas can and should articulate and what has to be carried over from one to the other to allow the psychiatrist to maintain his own sense of identity as he finds himself in what initially may seem like a strange new world.

In doing this we shall address ourselves to two main theoretical issues having a bearing on the problems mentioned earlier. The first pertains to the basic question of what we mean by the terms community psychiatry and social psychiatry. The second is concerned with exploring the parameters of the treatment process itself in psychiatry.

Community Psychiatry and Social Psychiatry

Each observer tends to define these terms idiosyncratically. We have previously alluded to the broad sense in which we use the term community psychiatry. The distinction between community psychiatry and social psychiatry can become clearer if we consider, for the moment, the two ways in which the term psychoanalysis is used. Psychoanalysis as a theoretical system concerns itself with the genesis of personality and personality disturbances. Psychoanalysis as a treatment system is concerned with the procedures and techniques necessary to ameliorate personality disorders. Analogously, we can view social psychiatry as a theoretical system, i.e., a system concerned with understanding the role of social determinants in mental illness. Community psychiatry then becomes applied social psychiatry.

Personality characteristics are a joint function of the person and his environment. A submissive person, i.e., will tend to remain submissive as long as the people around him can be seduced into dominating him. A neurosis, in other words, has to receive a nurturing feedback from the environment in order to endure. Each key social unit in which the patient is involved can become, in a sense, a "treatment unit,"[5] that is, it would have a certain therapeutic potential. Appropriate strategies of social therapy would have to be devised for these units beginning at the level of the family and extending to the neighborhood and the vocational milieu.

The Treatment Concept

This extended use of the term therapy certainly goes beyond ordinary usage of the term and raises the question as to whether or not such application does represent a logical extension of basic therapeutic concepts. In their paper, Rosenbaum and Zwerling delineate the qualitative differences between the individual and the social approach. They indicate the ways in which the traditional medical model is basically antithetical to the social psychiatric approach. Their resolution of the antithesis, by seeing the data as of a different order, thus implicitly suggesting a polite and mutual respect for the differences, does not quite get at the heart of the subject. We can agree with it as far as it goes, but it leaves out the elements of identity between the two endeavors. For these to emerge clearly it will be necessary to reassess the individual psychotherapeutic situation as a special case of the range of interaction with which community psychiatry concerns itself.

The tools which we have forged for changing behavior have arisen out of the dyadic relationship. When other techniques for changing behavior are judged by standards and goals arising in the one-to-one situation, we place ourselves at a disadvantage and, in a way, prejudicial to ourselves, misapply tools applicable in one setting to a different setting. Physician and patient are engaged in the common task of witnessing, identifying, understanding and changing self-defeating behavior, mostly in the patient, occasionally in the therapist. When the focus is on behavior it becomes readily apparent that part of the behavioral spectrum occurs outside the patient's awareness or his clear and objective apperception of it. Such behavior becomes diffracted psychologically so that its subjective reflection is a distorted and ineffectual one. Because of this the behavior in turn tends to remain outside of effective control and ultimately becomes a factor in symptom production.

The essential principles involved in the therapeutic approach are the structuring of a situation so as to elicit certain forms of behavior in the patient, the setting up of observer roles in the patient and therapist and the arrangement for therapeutic exchanges between the observers. This model, whereas it derives from the psychoanalytic situation and is directly applicable to it, nevertheless differs from certain features commonly identified with the psychoanalytic procedure. The concept of the patient operating from a regressed base is not intrinsic to this model. Nor is the concept of the therapist as a parent figure. We may have to revisit transference as Jackson[6] has suggested, bearing in mind that although these phenomenon have their roots in the past, the form and manner of their manifestation relate directly to the personality and activity of the therapist.

A model of this type begins with the objectification of behavior, assesses the concern with intrapsychic dynamics in individual therapy as intrinsic to an approach which relies mainly on verbal strategies in effecting healthier modes of adaptation and leaves open-ended the possible variety, value and importance of other ways of arriving at behavioral change. To recast the individual treatment situation in this way alters certain emphases, brings theory more in line with practice and allows for a clear identification of the lines of continuity between the small social unit, i.e., individual psychotherapy and the larger social units, the domain of social psychiatry.

It minimizes the occupational hazard of becoming so intrigued by intrapsychic dynamics that they become in effect, a tertium quid which keeps the operational field of the analyst within manageable bounds but may have the disadvantage of divorcing both him and the patient from certain social realities.

When individual therapy is interpolated at the level of social process it may be viewed as providing a microcosmic social laboratory in which the behavioral potential of the patient is examined through the interactional process set up between the patient and the doctor. The patient has to learn how to apply his analytic experience macrocosmically in his dealings with the world at large. The therapist, as a kind of midwife, facilitates this emotional emergence and, if successful, ultimately brings a new macrocosmic entity into existence.

In the broader based treatment approaches subsumed under the rubric of community psychiatry, the therapist or the therapeutic team learns first-hand something about the macrocosmic behavioral patterns of the patient and with this knowledge attempts to re-educate the patient to his own microcosmic identity. The patient discovers his real identity as a person as it is revealed in his dealings with another person. In psychoanalysis the therapist is, in the main, this other person. In community psychiatry, the patient's dealings with a broader range of people are brought under direct scrutiny.

When the psychiatrist moves from individual psychotherapy to community psychiatry he begins to gather data about behavior in a different way. He is still engaged, however, in the basic processes of inducing behavioral change, observing and interpreting the resulting behavior and inducing further change. In individual psychotherapy the focus upon the individual is intense and unrelenting. Psychodynamics emerge sharp and clear in the center of the stage. In the broader community approaches it is said that the dynamics are diluted and only tangentially identified. This leads to the invidious comparison referred to by Rosenbaum and Zwerling and others. What is lost sight of or downgraded, of course, are the manifold opportunities for behavioral change when treatment strategies evolve in connection with community resources and programs. The social psychiatrist is limited only by his imagination and energy in devising these social strategies, always granting that individual treatment approaches are necessary to consolidate the change and facilitate further change. Social psychiatry should add to the power of the therapist, not diminish it.

Summary

We have briefly discussed three phases of our own program. In each instance we are concerned with evoking, exploring and changing behavior in a social setting in a way that is as valid for the social psychiatrist as the same activity in the consultation rooms.

To effect a proper articulation between individual and social psychiatry, it may become necessary to take a hard look at some of the assumptions underlying individual psychotherapy and in particular, the extent to which we appear to have built into our theories a kind of psychological anesthesia inuring both patient and analyst to certain levels of social reality. Important theoretical insights are bound to emerge out of our current involvement in social and community psychiatry.

REFERENCES

1. Meyer, E., and Mendelson, M.: Psychiatric consultations with patients on medical and surgical wards: patterns and processes. Psychiatry 24:197-220, 1961.

2. YOCHELSON, L.: Sectarianism in psychiatry. Am. J. Psychiat. 120:1143-1148, 1964.
3. LEBENSOHN, Z. M.: A new role for the psychiatric unit of the community hospital. Comp. Psychiat. 4:375-360, 1963.
4. ROSENBAUM, M., AND ZWERLING, I.: Impact of social psychiatry. Arch. Gen. Psychiat. 11:31-39, 1964.
5. NADLER, E.: Personal communication.
6. JACKSON, D. D., AND HALEY, J.: Transference revisited. J. Nerv. & Ment. Dis. 137:363-371, 1963.

Discussion by Israel Zwerling, M.D.

THE DISCUSSION of theoretical issues by Dr. Ullman and his colleagues contains very lucid definitions of social psychiatry and community psychiatry as these terms have come increasingly to be understood; the statement, "Community psychiatry is applied social psychiatry" is likely to be widely quoted. However, the extended section on "The Treatment Concept" seems to reflect the pervasive hold of the psychoanalytic model on the authors. In effect, they are proposing an isomorphism between the patient-analyst and the patient-social group as two analogous dyads; the community psychiatrist, in their view, utilizes the latter widely, in addition to whatever use he may make of the former. The traditional analytic verbal transactions, focused upon the intrapsychic dynamics of the patient, are viewed as a two-person social system, and the patient's transactions with family, community and society are then seen as similar processes in more complex systems and in turn also subject to the therapeutic attention and perhaps intervention of the therapist in his efforts to achieve behavioral change in the patient. They state, "To recast the individual treatment situation in this way . . . allows for a clear identification of the lines of continuity between the small social unit, i.e., individual psychotherapy and the larger social units, the domain of social psychiatry." This represents, of course, a failure to recognize the different levels of integration represented by individuals and by groups, no less profound and no less subject to separate laws than are the properties or behavior of individual atoms and the properties of these atoms organized into molecules. What Dr. Ullman and his co-workers have in fact delineated is the continuum between individual psychoanalytic and individual "supportive" or "environmental" therapies. The *dis*continuity between the individual patient observed by an analyst and the social system observed by the social psychiatrist unfortunately remains with us. The most cursory perusal of the contemporary literature on the family, for example, will quickly make clear that the subject of study is no longer the fragmented elements and subgroups of the familiar "family history"—the schizophrenogenic mother or the passive father or the rivalrous sibling—described in the traditional individual case history. The subject of attention—research and therapy as well—is the family *per se*, a social system of forces and counterforces, unique and emergent, and not the sum of the individual psychic systems which are its elemental components—a system governed by laws, I would add, which so far have proved disconcertingly elusive.

THE "HERE AND NOW" AS THE FOCUS OF PSYCHOTHERAPY

By ALLEN J. ENELOW, M.D.
AND
LETA McKINNEY ADLER, Ph.D.

A GRADUAL SHIFT in emphasis from the content of patients' communications to the transactional process itself has characterized the development of psychotherapy. Freud was directing his attention away from the reconstruction and revival of early memories toward the immediately observable behavior of the patient when he wrote in 1914:

> "Finally there was evolved the consistent technique used today, in which the analyst gives up the attempt to bring a particular moment or problem into focus. He contents himself with studying whatever is present for the time being on the surface of the patient's mind, and he employs the art of interpretation mainly for the purpose of recognizing the resistances which appear there, and making them conscious to the patient."[8, p. 147]

Though Freud maintained his interest in uncovering forgotten "situations and connections," this particular directive to the analyst was the beginning of a long path away from "uncovering" as a psychotherapeutic technique.

As early as 1919, Karl Abraham[1, pp. 303-311] advocated that the therapist pay special attention to the way the patient was talking rather than immediately scrutinizing the content of his communications. Abraham was referring to what was then termed the "character-type." More specifically, he referred to the patient's way of behaving toward the therapist. In 1928, Wilhelm Reich[15] further developed this idea stating:

> "In addition to the dreams, slips of the tongue, and the rest of the communications of the patient, his attitude and behavior, that is to say, the manner

in which he relates his dreams, commits these slips, produces associations and communicates other material, deserves special attention." (p. 135)

Reich advocated scrutiny of the patient's manner of speech, the way the patient looks at and greets the analyst, the vocal inflection and every other detail of the patient's behavior in the consulting room. Reich added: "The *how* is just as important as *what* the patient says."

Hellmuth Kaiser, in 1934, while still considering "insight" the goal of therapy, proposed that the therapist's activity be *limited* to confrontations describing how the patient dealt with his (inferred) unconscious impulse as this could be observed in his overt behavior.[11] Unconscious impulses and motives were still to be inferred from the patient's free associations. Behavioral confrontations, Kaiser thought, would lead to insight about them, whether or not verbalized by the patient. This insight would be reflected in an increasing naturalness and richness of functioning characteristic of the healthy person. The bulk of the article was concerned with techniques for "resolving the resistances" through confrontation, with clinical examples.

Unfortunately, in this paper Kaiser gave a particularly vivid description of what he called a "true affective breakthrough." This referred to those moments when after analysis of resistance, a patient is able to express himself with considerable feeling and without artifice. Since Kaiser at that time conceived of such an event as representing a "freeing" of repressed instinct, it was mistakenly thought that his only interest was in promoting abreaction through analysis of resistance, and that he had no interest in so-called "working through." Strangely, Fenichel, usually so methodical, ignored the bulk of Kaiser's paper and criticized the ideas in it in these words: "Kaiser, an admirer of acting out, considers a 'true eruption of instinct' the sole therapeutic principle."[7, p. 107]

When Kaiser published again in 1955,[12] he proposed a psychotherapy in which no requirements were made of the patient except the single one of his presence in the consulting room. Kaiser's concern was to bring about in the patient an increased feeling of responsibility for his words and actions. Kaiser still maintained that the therapist's attention in this free and unstructured situation should be directed to the patient's behavior toward the therapist. The goal of therapy was no longer insight. It had become, instead, the bringing about of changes in the way the patient communicated. Any insight the patient developed was considered a by-product of the therapy. These ideas were further developed and discussed in Kaiser's third and last paper, "Emergency."[13]

In the meantime, other psychotherapists had been placing much more emphasis on the actual relationship between therapist and patient. In 1910, Freud had stated at Clark University that the transference which appears in every psychoanalytic treatment was based on no real relationship between the patient and the therapist.[9, p. 51] As the years went by, more analysts conceded that the personality of the therapist, as well as the forgotten feelings of the earlier years, had some determining effect on the relationship. Others began to emphasize, as did Clara Thompson, that one could view the experience of psychotherapy (psychoanalysis) as "an interpersonal situation where the ways in which the patient contributes to his current difficulties in living become manifest and clarified."[18, p. 220]

In 1948, Alexander's summary of his views de-emphasized interpretation of the transference and emphasized the corrective emotional experience occurring when the therapist's reactions to the patient differ from those of the persons (generally parent figures) with whom the conflicts are considered to be originally connected.[3, p. 276] Alexander suggested a planned regulation of the therapist's response to the patient in order to counteract the harmful effects of early parental attitudes. This proposal was contrary to Kaiser's view that consistently "genuine" behavior and communication on the part of the therapist eventually causes the patient's neurotic position vis-a-vis the therapist to alter or dissipate.

Other lines of development also led to an emphasis on the relationship between therapist and patient to the exclusion of reconstruction of the past. Some of the later formulations of Otto Rank gave impetus to the development of the relationship therapy of Taft and Allen. In *Psychotherapy with Children*, Allen wrote, "the therapeutic relationship is conceived as an immediate experience."[4, p. 49] Allen held that the patient will respond to the therapy situation in a manner characteristic of the problem for which help is sought and that the therapist must maintain his own integrity without becoming what the patient tries to make him.[4, p. 60] Allen made no effort to infer or interpret unconscious motives or reconstruct history.

In 1960, Enelow described the psychotherapy of patients who do not cooperate with the usual requirement that the patient talk to the therapist.[5] Following Kaiser, the therapist took the attitude that nothing was required of the patient other than his presence in the same room with the therapist in order for treatment to take place.[6] It was possible to demonstrate that effective therapy could be accomplished when the patient is silent much of the time.

Kaiser was unwilling to describe his activity with the patient except by clinical examples. He preferred to state that there were no rules for either patient or therapist. However, much of Kaiser's therapeutic behavior appeared to be in the direction of confronting the patient, either verbally or behaviorally, with what Kaiser called the "ungenuineness" in the patient's behavior. He later referred to this quality as "duplicity."

The present paper is an attempt to describe further what occurs in a psychotherapy—termed here "process-oriented" for ease of communication—which has as its philosophy the principles that: (1) the patient is required only to be in the presence of the therapist,* (2) that the therapist's interest be directed toward improving the ease, naturalness and spontaneity with which the patient communicates with the therapist, and (3) that the therapist respond to the patient in a spontaneous and natural manner. In our view, as in Kaiser's, such a philosophy of therapy tends to create a feeling of responsibility in the patient for his own words and actions.[12]

The process-oriented philosophy of psychotherapy is in line with Freud's emphasis on the importance of the patient's responses to the therapist. It also bears some resemblance to other therapies utilizing the "here and now" in therapy, those of Sullivan,[17] Alexander, Allen and others. It is evident, however, that its underlying philosophy is quite different. Although the patient's behavior is undoubtedly the same in substance as that characterizing similar relationships, present and past, the therapy situation is not used to understand and interpret the past in an attempt to modify behavior.

On the other hand, the therapist does not merely maintain a "good therapeutic attitude" in the vague hope that this will be therapeutically effective.[14] Nor is the therapist "nondirective" in Rogers' sense. Rogers holds that all that is required of the therapist for the occurrence of "constructive personality change" in the patient is the "communication

---

*The therapist may make other requirement such as payment of a fee. Though these may enhance the possibility of therapeutic success by contributing to the comfort of the therapist, they are not prescriptions for the patient's interview behavior nor for therapeutic success.

to the client of the therapist's empathic understanding and unconditional positive regard."[16] While the process-oriented therapist attempts to maintain a good therapeutic relationship and communicate his understanding to the patient, he is also actively engaged in a careful observation of the verbal and nonverbal communications of the patient and the formulation of interpretations and confrontations, both verbal and nonverbal, concerning the patient's overt, observable behavior.

It follows that the characteristic transactions of a process-oriented psychotherapy differ from those of therapies with other philosophies and goals. Once the patient is in the presence of the therapist, whatever the patient does or does not do is simply one more problem with which the psychotherapist must cope. The statement that the minimum possible requirements are made of the patient once he is in the presence of the therapist is meant literally. He need not talk and need not answer questions. He may say whatever he pleases and withhold any information he pleases.

The therapist primarily concerns himself with the way the patient behaves in the session. The way he talks or does not talk is chiefly the subject of what the therapist says to the patient. He behaves in such a way as to bring to the attention of the patient (by verbal or by behavioral confrontation) the elements of his communicative behavior toward the therapist that are incongruous, inconsistent, strike a false note, or would serve (if successful) to bring the therapist into an unhealthy kind of interaction with the patient; e.g., dominating, being dominated, being treated as incompetent or helpless, getting involved in altercations, etc. His interpretations are uniformly characterized by a low level of inference with respect to past or present intrapsychic phenomena. Suggestions, verbal reassurance, and advice are not given. The therapist avoids such standard devices as asking, "Could you tell me more about that?" or "What were your thoughts about that?" as well as some of the other subtler ways of indicating special interest or disinterest in certain topics rather than others.

In this process, the therapist is engaged as a real personality. While he makes no effort to preserve anonymity, he acts in the role of physician as defined by our culture—a helping, somewhat permissive figure, invested at the same time with the authority of the expert who maintains those limits in the reciprocation of responses to patients implied in all professional relationships. This is in contrast to the view, held by Ackerman and others, that the psychotherapist is an "illusory image which changes from time to time as the patient projects his fantasies on the person of the analyst," and who does not provide "a genuine social experience to the patient."[2]

What happens between the therapist and the patient, if one works in this way, appears to be something like this: The patient behaves in the way that characterizes his relationships with others, certain peculiarities of his behavior being somewhat accentuated by the very fact that he looks for a great deal from the therapist. The therapist, by being himself, does not enter such a relationship; at the same time, he makes the patient aware of the impact of the patient's behavior on him. When treatment succeeds, the patient's way of communicating with the therapist changes over the months of talking about it. Concomitantly, a gradual amelioration of symptoms occurs even though they may rarely if ever be mentioned.

Two Examples of Process-Oriented Psychotherapy

Obviously, since the philosophy of this psychotherapy includes the view that the therapist should respond genuinely and spontaneously, the two examples which follow illustrate what happens when a particular therapist encounters a particular patient. Other therapists and other patients will interact differently even though the philosophy of therapy is the same as that described here. While fragments of the interview sessions are quoted from the therapist's memory, they are reported with sufficient accuracy to give the authentic flavor of the interaction.

Mrs. Shirley T. was a 32-year-old married woman referred to the therapist by her family doctor because of an intractable peptic ulcer. In the initial interview, the patient talked at some length about her present situation and her history. She was one of two children. She had been raised by a nursemaid and hardly knew her own parents. When she was with them, however, there was much demonstration of affection in a rather artificial way.

She married at a relatively early age and had two children. Her husband was a professional man who recently had suffered a heart attack. He was somewhat undemonstrative but had been ordinarily attentive until his heart attack, after which he became extremely inattentive, self-centered, and irascible. The patient had had, under conditions of stress in the past, two transitory peptic ulcers which responded well to medical management. Now, however, her ulcer continued to be active and symptomatic despite months of treatment. She became depressed, indecisive, and preoccupied with her husband's behavior.

The most remarkable thing about the patient in the initial interview was that she recited the many wrongs done to her by her husband with an absolutely expressionless face. Even more noticeable was the fact that when she smiled, her mouth did the smiling, or when she wept, her eyes did the weeping. Her general facial musculature took no part, producing a remarkable wooden and unconvincing expression. Her account indicated that she felt no internal difficulty (other than a stomach ache). Her problems, as she saw them, were thrust upon her by a disturbed and disturbing husband.

Her second interview began:

Patient: What is it that I am expected to do here?

Therapist: (After a moment) Nothing, really, if I understand your question. It sounds like you are saying: "What do you want me to do?" There's nothing that I want you to do.

She looked puzzled. After a moment, she shrugged as though to say, non-verbally, "What a strange man this is," and began to talk about what had happened since her previous appointment. This consisted of a detailed account of her husband's self-centered behavior.

After about twenty minutes, the therapist became so aware of the curious woodenness of expression which had occurred in the first interview that he could hardly concentrate on the patient's words.

Therapist: You know, it is a very interesting thing that when you smile, nothing takes part in the smile except your mouth.

The patient looked uncomprehending, but after a moment, shrugged and went on talking. Soon tears came to her eyes while her face remained unchanged.

Therapist: I noticed that you were crying just then, but nothing moved in your face while you cried.

Patient: (Irritated) What do you mean by that?

Therapist: Just what I said. It is a kind of odd sensation to speak with someone whose face shows as little feeling as yours does, even when you seem to feel something strongly.

Patient: But, I don't understand. What am I expected to do with that?

Therapist: Nothing, really, I guess. I just wanted you to know how curious this looked to me.

This might be said to characterize the interaction between patient and therapist throughout the therapy. She never seemed really to understand just what point the therapist intended to make, but there was a progressive change in her behavior.

There was a remarkable similarity of theme whenever she spoke: complaint about the hardship she suffered because of the unreasonable behavior of another. At first this centered mainly on the self-centered churlishness of her husband. Sometimes it was her mother's unwillingness to inconvenience herself for the patient. As time went on, her husband became less often the subject of the complaint. More often it was the husband's partner, her in-laws, her son, or the wives of her husband's colleagues. From time to time, the therapist commented on this, taking one of the following points of view: (a) "You speak as though everything that happened to you was at the instigation (or due to the pressure) of someone else," or (b) "Curious—today it is your mother-in-law. Last week it was your husband's partner. But the subject really remains the same."

At times the tone of her voice and her general manner were those of someone who performs a task out of a sense of duty but without real interest in it. That is, she spoke as though she were *required* to come to the office, *required* to speak and dully, even boredly, pushed herself through it. The therapist described this to her. At first, this annoyed her; later, she was amused. Gradually, the stiffness in her manner changed. Her general appearance became considerably more animated. More and more she spoke of herself and how *she* felt, less and less about the unfairness of the behavior of others.

After about fifteen months of treatment, the patient came in one day and said, "I feel very much better. I have not had stomach pains in more than five or six months." (This was the therapist's first information that she no longer had ulcer symptoms.) "I cannot explain what it was that did it. I talked it over with my husband and neither of us can understand why these silly inconsequential things that you say could possibly have helped me. But whatever you did, it was helpful. I feel like a different person and everybody says I've changed in many ways. My husband says I am more human than I used to be. So, thank you very much. I won't be coming any more after this week."

An opportunity for an accidental followup occurred some two years later when the therapist met the patient and her husband at a party. She greeted the therapist with: "I can never thank you enough for what you did for me even though I don't understand how you did it. I have been feeling fine ever since."

The second case example is taken from the treatment of Mr. K. C., a 37-year-old, married, actor.

He played minor roles in television shows, usually a "heavy"—a cruel person, often a criminal. In contrast, he was actually a mild person, often too shy to eat in public places or to attend parties. He was hungry for affection and fearful of being alone. He had frequent anxiety attacks and constant hypochondriacal preoccupation. Headaches made him fear a brain tumor, upset stomachs filled him with apprehensions about cancer. He consulted the senior author after an anxiety attack in which he was unable to go on with a scene. He could not remember his lines, was fearful of speaking and feigned a "fainting spell."

In treatment he found it difficult to speak without some immediate response from the therapist. Thus he would begin an account of a recent occurrence, but, within a few minutes, would begin to slow down while anxiously scrutinizing the therapist's face. "Do you understand what I mean?" he would ask. If the therapist said "Yes" or nodded, he would sigh, as if in relief, and go on.

After it occurred several times, the therapist described this to the patient. Mr. C. looked thoughtful, "Why do you think I do that?" he asked. The therapist said, "I don't know."

This behavior continued to dominate the early hours. On another occasion when this occurred, the following interchange took place:

Therapist: When you are talking, it isn't more than a few minutes before you examine my face for a response, as though you needed to know at every moment if your message is being received.

Patient: I know. You've said that before. But what should I do about it?

Therapist: You mean other than having psychotherapy?

Patient: (Smiles) I see. You feel I'm already doing something about it.

Several weeks later, the same issue came up again as follows:

Therapist: Sometimes you give me the impression of a man who cannot see the person with whom he tries to make contact. Sort of like a short-wave radio operator who sends out messages without knowing if they are being received; so from time to time he says something like: "Do you read me? Come in, please. Over."

Patient: (Smiles as if in recognition) I have a feeling just about like that whenever I talk to anybody. I feel nervous and I wonder if they're really paying attention. I guess I want everyone to reassure me—to say "Yes, yes, we care about what you're saying."

Another theme in the treatment of this patient was his frequent requests of the therapist to tell him what he should feel about a certain thing:

Patient: Then he turned to me and said, "You have a great deal of talent but you have done nothing to promote yourself." I didn't know what I should say or how to feel. Tell me, what should I feel in that kind of situation?

Therapist: It's striking! You say, "What should I feel" as though you had no feeling of your own.

Patient: I could have feelings if you would only tell me what they should be.

Therapist: Like your director does? I feel that you'd like me to be a sort of psychological director for you.

Patient: (Sadly) It would be very nice. Too bad you won't.

This patient gradually became much more free in his expression. Within eight or nine months after treatment began, he no longer anxiously inquired if the therapist were still "reading him."

Mr. C., after three years of psychotherapy, was free of anxiety attacks and working more comfortably than he ever had before. Because of his increasing ability to express himself freely and to enjoy social situations, he was working more regularly. He also showed greater ability to relate to his wife as one adult to another where formerly he had been attached to her in an infantile clinging way.

The Place of Process-Oriented Psychotherapy In a General Practice of Psychiatry

To give an idea of where process-oriented psychotherapy has application in a general psychiatric practice, some gross tabulations based on the total number of patients who entered treatment (excluding evaluations and consultations) with the psychiatrist-author from January 1, 1956, through December 31, 1961, are presented here. These 133 patients represent about 35 percent of all patients seen for the first time during the sample period. The proportion of patients treated remained almost constant from year to year during the sample period. Among the treated patients, 47 per cent received process-oriented psychotherapy exclusively; an additional 12 per cent received process-oriented psychotherapy modified by drugs and/or by an increase in supportive behavior; seven percent received supportive therapy without drugs, while 23 per cent received drugs and supportive or directive therapy; and 11 per cent received ECT in conjunction with some form of psychotherapy (table 1).

These categories of treatment are based on the concept of a spectrum of psychotherapies. At one end is reconstructive psychotherapy in which the therapist endeavors to change the nature of the patient's relationships. At the other end is supportive psychotherapy which does not set that goal though such changes may occur. In all therapies, the active interest or concern of the therapist is felt by patients as supportive. In reconstructive therapy, the supportive elements are minimized; in supportive therapy confrontations concerning the patient's way of communicating are minimized. Throughout the spectrum, the process-oriented therapist directs his attention to what goes on in the "here and now"; i.e., the transactional process itself.

Psychoanalysts have made fairly explicit the position that psychoanalysis is primarily appropriate for persons of given age ranges and certain symptomatologies. No comparable qualifications for process-oriented psychotherapy have been established. The only consciously held basis for selection has been that the therapist not dislike the prospective patient (and that the total practice of the therapist pay him adequately for his work.) In table 1, the treatment received by patients in different age groups is shown. Whereas 58 per cent of the patients forty-five years of age or under received process-oriented psychotherapy exclusively, only 18 per cent of those over forty-five years of age received this treatment.* Thus, the selection of patients by age for process-oriented therapy was similar in practice to the principle on

*This difference is significant at the .001 level by the chi-square test with one degree of freedom and corrected for continuity.

which psychoanalytic patients are selected. Since there are no known comparable data for psychoanalytic practice, it is not possible to say whether or not the 7 of 62 patients over 45 years of age who received only process-oriented therapy represent some relaxation of age limitations.

TABLE 1.—*Type of Treatment by Age of Patient*

Type of Therapy	25 Years or Under	26-35 Years	36-45 Years	46-55 Years	55 Years or Over	Total (Number)	Total (Per cent)
Process-oriented psychotherapy	8	25	22	6	1	62	47
Modified process-oriented therapy*	2	1	10	3	0	16	12
Supportive therapy without drugs	2	1	3	3	0	9	7
Supportive and/or directive therapy with drugs	2	4	9	4	12	31	23
ECT with support or directive therapy	0	2	3	5	5	15	11
Total Patients	14	33	47	21	18	133	100

*Six patients received drugs and seven received supportive therapy along with process-oriented therapy. Three patients were started in process-oriented therapy but were shifted to supportive therapy, one with drugs as well. The three groups were similarly distributed by age.

In table 2 are shown the kinds of treatment received by patients classified into major diagnostic categories. More than eight in ten patients diagnosed as having a personality disorder, a psychophysiological reaction, or an obsessive compulsive reaction were treated with purely process-oriented psychotherapy. Two-thirds of the depressive reactions and one half the anxiety reactions were similarly treated. About one in four of the patients diagnosed as alcoholic, schizophrenic, or manic-depressive were treated by process-oriented psychotherapy exclusively, whereas it was not the treatment of choice for the involutional psychoses.

Recently, some behavioral scientists have raised a question concerning the appropriateness of insight-aiming therapies for manual workers, pointing out that faith in the solution of problems through understanding and knowledge is more characteristic of the middle and upper classes. Some investigators have asserted that the middle-class, insight-aiming, psychodynamically oriented psychotherapist both rejects and is rejected by lower-class patients.[10] In this practice, one-half the patients coming from a household in which the head was either an executive or a professional person were treated exclusively by process-oriented psychotherapy, two of five patients coming from other "white collar" families and an equal proportion from "blue collar" families were so treated (table 3). This does not represent a statistically significant difference by occupational status.

TABLE 2.—*Type of Treatment by Diagnostic Category*

Diagnostic Category*	Process-Oriented Psychotherapy	Modified Process-Oriented Therapy	Other Therapy	Total
Personality disorder (not including sociopathic disorders)	16	0	2	18
Psychophysiologic disorder	6	1	0	7
Obsessive compulsive reaction	4	1	0	5
Depressive reaction	19	2	8	29
Anxiety reaction	6	4	2	12
Alcoholism	4	2	9	15
Schizophrenic reaction	4	2	10	16
Manic depressive reaction	2	2	4	8
Involutional psychosis	1	2	20	23
Total patients	62	16	55	133

*Ordered according to the proportion of patients with the given diagnosis who received process-oriented psychotherapy exclusively.

TABLE 3.—*Type of Treatment by Occupational Status*

Occupation of Head of Household	Process-Oriented Psycho-therapy	Modified Process-Oriented Therapy	Other Therapy	Total
Professional and Executive	41	10	31	82
Other white-collar	16	6	17	39
Manual workers	5	0	7	12
Total patients	62	16	55	133

SUMMARY AND CONCLUSIONS

Freud's interest in "the transference" and his remarks calling attention to the need for examining the "surface of the patient's mind" during the therapy session have been elaborated in several different directions by later psychoanalysts who emphasize the "here and now" in psychotherapy. Several of these movements have been discussed in this paper. They have been described in order to differentiate one of these, termed "process-oriented" psychotherapy, from the others. This psychotherapy had its origin in Kaiser's extension of Reich's proposals to the effect that the patient's way of communicating with the therapist should be the only material interpreted by the therapist. It became, largely through Kaiser's subsequent development, a therapy differentiated from other therapies of psychoanalytic origin in four major ways.

(1) Nothing but his presence with the therapist is required of the patient. This quality of the therapy is essential because any attempt on the therapist's part to direct the patient's behavior during the therapy session interferes with the patient's characteristic modes of interaction, and it is precisely these with which the therapist deals. These modes of interaction usually are manifested in an attempt to involve the therapist in some form of neurotic relationship. When nothing is required of the patient, the particular kind of control he seeks from the therapist is withheld.

(2) The therapist offers a spontaneous, natural, nonmanipulative relationship as a means of avoiding involvement in the kind of neurotic

relationship the patient seeks to establish as a result of his psychiatric disturbance.

(3) The comments of the therapist tend to deal with the patient's behavior during the therapy hour rather than the content of the patient's communication *per se*.

(4) Interpretations are of a very low order of inference both with respect to causation and motivation. Inferences from specific observable behaviors usually concern the underlying transactional process between the patient and the therapist of which the observed behavior is considered to constitute a fragment.

Since the therapist with this philosophy behaves in accordance with his own personality and in response to a patient whose behavior is not stereotyped by the therapist's requirements, each therapy is unique. To illustrate the kinds of interventions made by one therapist who follows this treatment philosophy, excerpts from the therapy of two patients were presented.

In order to provide some objective data about the place of process-oriented therapy in the private practice of general psychiatry, data concerning the type of patient selected for this therapy were presented. Almost one-half of all patients treated over a six-year period were so treated. Process-oriented therapy was modified by increased supportiveness and directiveness and by pharmacological treatment, but the focus of all therapeutic interaction was the doctor-patient transaction. There was a tendency to select younger patients and certain diagnoses more frequently for reconstructive process-oriented therapy. The occupational status of the family head was not a statistically significant factor in the mode of treatment selected.

REFERENCES

1. Abraham, K.: A particular form of neurotic resistance against the psychoanalytic method. *In* Selected Papers of Karl Abraham, E. Jones, ed. London, The Hogarth Press, 1942.
2. Ackerman, N. W.: Social role and total personality. Am. J. Orthopsychiat. 21:1-17, 1951.
3. Alexander, F.: Fundamentals of Psychoanalysis. New York, W. W. Norton, 1948.
4. Allen, F. H.: Psychotherapy with Children. New York, W. W. Norton, 1942.
5. Enelow, A. J.: The silent patient. Psychiat. 23:153-158, 1960.
6. ——, and Adler, L. M.: What is Required of the Patient in Psychotherapy. University of Southern California, School of Medicine, Department of Psychiatry (mimeo.).

7. FENICHEL, C.: Problems of psychoanalytic technique. Albany, The Psychoanalytic Quarterly, Inc., 1941.
8. FREUD, S.: Remembering, Repeating and Working Through. *In* The Complete Psychological Works of Sigmund Freud, vol. XII. London, The Hogarth Press 1958.
9. ——: Five Lectures on Psycho-Analysis. *In* The Complete Psychological Works of Sigmund Freud, vol. XI. London, The Hogarth Press, 1957.
10. HOLLINGSHEAD, A., AND REDLICH, F. C.: Socioeconomic Status and Mental Illness. New York, Wiley, 1958, p. 192.
11. KAISER, H.: Problems of technique. Internationale Zeitschrift fuer Psychoanalyse 20:490-522, 1934.
12. ——: The problem of responsibility in psychotherapy. Psychiat. 18:205-211, 1955.
13. ——: Emergency. Psychiatry. 25:97-118, 1962.
14. LEVINE, M.: Principles of psychiatric treatment. *In* Dynamic Psychiatry, F. Alexander and M. Ross, eds. Chicago, University of Chicago Press, 1957, p. 357.
15. REICH, W.: On character analysis. *In* The Psychoanalytic Reader, R. Fliess, ed. New York, International Universities Press, 1948, p. 135. Translated from the original Ueber Charakteranalyse. Internationale Zeitschrift fuer Psychoanalyse 14:180, 1928.
16. ROGERS, C. R.: The necessary and sufficient conditions of therapeutic personality change. J. Consult. Psychol. 21:95-103, 1957.
17. SULLIVAN, H. S.: Conceptions of Modern Psychiatry. New York, Norton, 1948.
18. THOMPSON, CLARA: Psychoanalysis: Evolution and Development. New York, Hermitage House, 1950.

Discussion by Marianne H. Eckardt, M.D.

Alan Watts writes in the introduction to his new book on myths: "When the critical intellect looks at anything carefully, it vanishes. This is true of the solid substance of bodies as of historical generalizations, of entities such as nations, of epochs such as the Middle Ages and of subject matter such as myth. The reason is, of course, that "things" exist only relatively—for a point of view or for convenience of description. Thus when we inspect any unit more closely we find that its structure is more complex and more differentiated than we had supposed. Its variety comes to impress us more than its unity. This is why there is something of the spirit of debunking in all scholarship and scientific inquiry."[1]

Alan Watts in these few sentences summarizes what in essence I will be saying in my discussion, and the spirit of debunking is, of course, also present. I believe that discussion of psychotherapeutic technique, which is very much needed, is profitable only in the context of either our vision of a specific clinical problem or our idea of psychoneurotic development and functioning, preferably with frank acknowledgment that effective technique is complex indeed and very differentiated.

The moment a psychotherapy gives itself personal identity by focussing on a special technique and by adopting a trademark name, it is asking for trouble. Vested interests enter and the freedom of openminded development is curtailed. The main problem, however, arises because the original rationale of the technique, which infused it with life and justification, becomes lost due to generalization.

Freud's technique of free association belonged to his theory of repression, and his search for a way into the unconscious. As soon as theories shifted for instance to a different concept of ego function other techniques entered.

Dr. Enelow wrote a beautiful paper on "The Silent Patient"[2] in 1960. It arose in part out of the immediate influence of Dr. Kaiser, in part out of his own frustrations in treating patients who did not cooperate in telling what was on their minds. He was aware that his expectations of the patient to talk engaged him in a power struggle and rather logically he thought of removing his side of the struggle. The patient was told that nothing was expected of him and Dr. Enelow limited himself to statements about observable behavior on the patient's part. The paper describes the dissolution of this obstacle and the improvement of the patient. In this paper frustration in treatment, a new view of the dynamics of the situation, and the action taken blend into a logical sequence. In so far as other therapists encounter a similar situation they might well benefit by Dr. Enelow's experience.

Dr. Hellmuth Kaiser, as reflected in his all too few papers, must have been an outstanding person as well as an outstanding therapist. His papers, about technique, are excellent. He was the first to formulate the technical statement that nothing was expected of the patient but to be in the presence of the therapist. This formulation arose out of his concern with responsibility both on the part of the therapist as well as of the patient. He was displeased with the therapists' tendencies to blame patients for failure of therapy rather than to face up to the fact that either there was something wrong with their diagnosis and prognosis for therapy or with their approach. His main concern, however, belonged to what he felt was a general phenomenon of neurosis, namely that patients did not seem to stand behind their words, that in this sense they did not feel responsible for their feelings and actions. He felt the patient was able to hide behind the rules of the therapist. Hence, his dictum that nothing was expected of the patient but his presence with the therapist. Dr. Kaiser's aim thus was to expose the patient's tendencies to hide behind rules and imagined shrouds, and to lead him to stand behind his words.[3] Here again we meet the alive, meaningful connection between the frustration with an old procedure, the specific vista of the patient's dynamics and the imminent logical action.

The chapter by Enelow and Adler still retains some of Dr. Kaiser's reasoning, but, I feel, in much weakened form. Their main emphasis is no longer to put the patient behind his words, but the observation of the patient's "characteristic modes of interaction" and it is these, they feel, with which the therapist ought to concern himself. They stress their desire not to lend themselves to the patient's neurotic entanglements and stress the importance of avoiding inferences about the dynamics of their behavior and historical development. The dictum of nonexpectation aims thus chiefly at preventing neurotic misuse of more active therapeutic involvement. This contrasts markedly with Dr. Kaiser's presentation as seen in his brilliant article "Emergency."[4] Here it becomes clear that for him the technical step of nonexpectation meant to challenge, expose, at time provoke and disconcert the patient. It was to him a very active measure in intent. Dr. Kaiser dared the unorthodox; he made a point of being all there as a person. In his fictionalized case history he reacts to content and manner; he questions, philosophizes, and knows there were risks to be taken in treatment and no safety in one procedure. The present authors are much more concerned than Kaiser was with not injecting themselves into the situation. I wonder whether they do not underestimate the degree to which our biases, morals and judgment enter into all our statements, even those we consider to be only observational.

Kaiser's emphasis on genuineness and presence is well brought out in a passage of dialogue in his article "Emergency." The fictional dialogue is between two therapists, one of whom is, unbeknown to himself, the patient. "Look," the therapist says, who is the spokesman for Kaiser, "it is a sad truth that rarely, very, very rarely do people say what they feel like saying. Here we are, the both of us, in this office together, free for an hour to say what we think. We do not have to sell anything to each other, nor do we have to agree on by-laws or resolutions. We don't have to fight each other, beguile each other, persuade each other. We might not always grasp immediately the other's meaning, but we have the potential of doing so. Why waste this unique opportunity?" He was asking for whole-hearted investment of the therapist in the process and not for limitations and exclusions.

REFERENCES

1. Watts, A. W.: The Two Hands of God. New York, George Braziller, 1963, p. 1.
2. Enelow, A. J.: The silent patient. Psychiatry 23:153, 1960.
3. Kaiser, H.: The problem of responsibility in psychotherapy. Psychiatry 18:205, 1955.
4. ——: Emergency. Psychiatry 25:97, 1962.

A THERAPEUTIC NURSERY GROUP IN A SETTLEMENT HOUSE

By SHERMAN O. SCHACHTER, M.D. AND
DAVID L. WOLITZKY, PH.D.

THERE ARE two major population groups which have benefitted little from all the advances made in the field of mental health. These two groups are the very poor and the very young. Insofar as the poor are concerned, the possibility of getting help for mental illness goes down as social status goes down.[1] A patient's social status can determine whether he will be accepted at a clinic, the length of his treatment, the kind of treatment and even which staff member will treat him.[2] Moreover, if a lower-class patient does start therapy, he and his middle-class therapist communicate only with great difficulty.[3,4,15] Paradoxically, while poverty increases the incidence and severity of mental illness, at the same time it decreases the possibility for help. The resignation, helplessness, ignorance and misconceptions which acccompany the lower economic status also serve to frustrate solutions of emotional problems. Frequently there is a lack of awareness among the poor of what constitutes mental illness as well as an ignorance of mental health resources in the community. Someone struggling to feed a family, or worried about the rent week after week, or concerned about the loss of work or the absence of any chance to work, is not likely to be sensitive to deviant development in a very young child. Even more, under these conditions, a severe degree of inhibition or withdrawal in a child may actually be welcomed.

A psychiatric clinic in a settlement house, with a patient population drawn from the underprivileged, and with many of these patients preschoolers, is faced with these problems.[14]

The Hudson Guild Neighborhood House is a settlement house which has been part of the Chelsea community of New York for more than seventy years. As times changed and our social structure changed, so did the neighborhood, and with it the concept of social service. In 1943, the Settlement House established a children's Day Care Center, an all day nursery for the children of working mothers and others who need nine to five care for preschool children. As would be expected, many of these children were found to have severe emotional problems. In 1948, in response to requests for help from the children's Center and in response to the community need for help, a small psychiatric clinic was set up in the Settlement House.

As the demand for treatment grew for these preschool children, an experiment was tried. Since many of the development tasks of this four- to five-year-old-age group are in the area of social adjustment, what was needed was a therapeutic approach in a group setting. A therapeutic nursery group (TNG) with a teacher-therapist, was established. TNG is a treatment group of four- to five-year-olds, about five in number, who are taken from their regular Day Care Center class to meet twice a week, an hour each session, with their teacher-therapist. It is this unique therapeutic group, in operation for the last eight years, that we wish to discuss in this paper. When the group was first reported as a pilot project by Lambert, Mahler and Moore,[5] their report was enthusiastic, and we are equally enthusiastic today. We shall first describe the group briefly and then discuss its implications for community mental health.

Description of TNG

The Day Care Center, from which the TNG children are selected, is located in the midst of a low income housing project. The Center is operated by Hudson Guild and supported by funds from the New York City Department of Welfare and Hudson Guild.

Our children come from deprived families, mostly Negro and Puerto Rican. Fathers are frequently absent from the home; if present, they are unskilled, occasionally semiskilled, laborers. Periods of unemployment are frequent. Most mothers work and some families receive relief. In general the families are culturally and socially impoverished. Often the family speaks Spanish at home and the children have difficulty with English. In this culture of poverty, family life is frequently unstable. The number of children in several of our families exceeds six. Separation, divorce, desertion, illegitimacy, are common. This is the atmosphere in which the problems of these children flourish.

The first step in the selection of children for TNG is made by the regular Day Care Center teacher who, through her own training and experience and her consultations with the TNG personnel, is particularly sensitive to deviant behavior.

Some of the symptoms which lead to selection in the four- or five-year-olds include withdrawal, immaturity, aggressiveness, fearfulness, overdependence, incontinence, temper tantrums, hyperactivity, isolation from peers and mutism. To choose the children who pose the greatest problems for the teacher would be an easier means of selection; however we'd simply have a group consisting of aggressive brawlers. Rather, these are children who show signs and symptoms of many varied psychic conditions. Because they are in an all-day living situation, difficulties can be spotted even though they may have no nuisance value. For instance, Rita spoke only to mother and father and to no other adults. For two years in the Day Care Center she would not address or answer the teachers. She avoided even passing close to them. Yet at the same time she was cheerful, cooperative, played well with other children and was liked by them. More typical is Carl—frequently moved from country to country, home to home, relative to relative; he was aggressive, destructive and a terror to family and teachers alike.

When a teacher presents the name of a prospective patient to the Director of the Day Care Center, we ask the family for permission to study the child, even though the family has not come to us with any complaint. There is a rather detailed diagnostic process which includes a history-taking interview with the parents by the clinic social worker, psychological testing and psychiatric evaluation of the child, and observation in the Day Care Center class by the TNG teacher-therapist. In early summer, there is a conference held for selection of children for next year's group. Meeting together are the Medical Director of the Clinic, the psychiatrist who examined the child, the social worker who interviewed the parents, the psychologist who tested the child, the TNG teacher-therapist who observed the child, the Day Care Center teacher who referred the child, and the Director of the Day Care Center. At this meeting, a tentative dynamic formulation is reached and a specific treatment plan for both child and parents is developed. If we do not accept a child for TNG, an alternative plan of treatment is offered.

There are two nursery groups of five to eight, one at approximately the four-year level and one about the five-year level. At times, the older group will contain a child who has not gone on to the first grade but is being given another year of treatment and nursery school. That number seems to allow for enough group interaction while still insuring adequate attention from the teacher-therapist. The groups are started in September and run through June. They meet twice a week in a playroom in the Day Care Center for hourly sessions, with a goal of approximately eighty sessions per year.

The teacher-therapists are trained in nursery school education and have a good working knowledge of clinical diagnosis and therapeutic theory and technique. They are nursery school teachers and have not had previous experience in play therapy. Many of their treatment techniques were evolved by trial and error in conjunction with frequent discussions and psychiatric supervision. Our teacher-therapists have been trained at the Hudson Guild for their task.*

The teacher-therapist intervenes in specific ways. For example, Hector was four and a half years old, immature, disruptive in his behavior, superficial in all his relationships, and ate almost constantly. There was a history of repeated or prolonged separations from his mother. In a group, when anxious, he would wander aimlessly with a sad forlorn expression and interfere with other children. At such times the teacher-therapist would intervene and supply physical closeness by holding him on her lap. After several months, Hector was able to make meaningful contact with other children and engaged in cooperative, nondestructive play.

The therapist's attitude is determined by the treatment goals for each child. Jamie, who functioned as a pseudodefective, needed constant praise and encouragement and support. Yvette, who controlled everybody around her through her brilliant manipulations, needed help in accepting less than perfection. Marco, tense, angry, aggressive and demanding was helped to verbalize his anger with other children rather than attack. He started making friends for the first time.

With many of these severely damaged and traumatized children we have found that if we can remove a major stumbling block to development, we may leave much in the child that is still ill, but we permit development to continue with less distortion. Robert was a case of a strange, restless, anxious, silly boy of five, racked by anxiety. His thinking was illogical and he heard the devil's voice. He made no contact with the other children and did not relate to the teacher-therapist. Through a stuffed kitten given to him by the teacher-therapist he was able to make tentative contact. The kitten was first a monster, later very sick, and still later, playful. At the end of the

*Our original teacher-therapist was Mrs. Frances Schor. A short while later Mrs. Mary Jane Wittenberg joined her. It is only through their dedication that the program was made possible.

school year the kitten was discarded and Robert was playing with other children, expressing himself clearly and much less fearfully.

An effort is made to involve each parent in a counseling relationship with our clinic social worker. These parents do not come to us asking our help for their child; we come to them, explaining that something is wrong and that we want to help. The degree to which we involve these parents depends in great part on the skill of the worker in dealing with initially unmotivated parents who are not very verbal or sophisticated. In one of the few studies reporting successful psychotherapy *not* being related to social class[12] the significant statement is made ". . . we are more concerned with nurturing, than with assessing, their motivation." Some families need supportive help in their chaotic, disintegrating life situations. With some parents discussions of their marital relations takes immediate pressure off the child who is bearing the brunt of a difficult marriage. With other parents, uncovering psychotherapy can be employed.

Implications for Community Mental Health

In a neighborhood of high delinquency, gang activity and crime, *none* of our TNG graduates—and the oldest are twelve and thirteen—has been in trouble with the courts. We find this encouraging.

Ours is not a therapeutic nursery, but a therapeutic group within a normal nursery.[9,13] Because it is so simple, it could be duplicated by any day care center. Our Day Care Center happens to be connected with a Settlement House but the Therapeutic Nursery Group could work without that. It would simply require personnel to be portable and transported to the centers. There are many programs for detection and diagnosis of primary grade children, but very few for preschoolers. The successful functioning of a therapeutic nursery group in preschool setting would provide early detection and early intervention and should serve to reduce the subsequent severity of disturbance. In New York City alone, there are over eighty day care centers with about six thousand children. A great many of these children are headed for serious disturbance. The New York City school system is taking some initial steps in the area of detection by setting up prekindergarten classes for children from deprived neighborhoods. There is an increasing number of reports of psychiatrists spending consultative time in normal nurseries[13] but this is not enough since those needing referral for treatment would too often simply have no place to go. Early intervention is very worthwhile, but is very much neglected.

On the Use of Group Treatment

Our experience has emphasized for us the effectiveness of group treatment for this age group from this population. Group treatment is particularly appropriate because of the presence of what Slavson[10] has called "social hunger." The do's and don'ts of technique are fully discussed in a recent volume dealing with the "unique contributions of group play therapy."[11] This text gives many criteria for not accepting children in groups—criteria which we do not follow. Since our resources are limited, the opportunity to treat several children simultaneously is welcome.

On Reaching the Parents

Burdened by the everyday demands of their hard struggle to clothe and feed, ignorant of mental health facilities or activities or services, suspicious of agencies, and unable to recognize serious pathology in a child, our parents *don't* come to the clinic. But, remarkably, when approached by people they already like and trust (from the Day Care Center) in a setting which is familiar and being helpful to them (the Settlement House), they respond and cooperate with surprising willingness. A few parents respond because of their fear of authority, and remain sullen, grudging and covertly aggressive, but most respond with sincere interest in helping their child. Neubauer[8] in an editorial commenting on the neglect of the preschool child by clinics and child guidance agencies, sums it up by saying "we must move from giving services only to those who ask for help to helping those who need it."

On the Use of Teacher-Therapists

We have found that the nursery school teacher is well suited for the group treatment of preschool children. When one of our teachers left last year after seven years as a teacher-therapist, we developed an on-going training program. There was much discussion at the time as to whether we should train nursery school teachers to do this kind of therapy, or teach nursery school techniques to group therapists. We felt that the training of teachers was more appropriate: they are more available; they are well versed in dealing with preschool children; they already use many techniques which are essentially therapeutic, though not labeled as such. Many educative techniques are appropriate to help children of this age master new tasks. That part of therapy which is learning, increases as the age of the patient decreases. For four- and five-year-olds, to teach and treat, have very much in common.[6] It is this that allows a nursery teacher to develop more easily into this special kind of therapist. There are not enough workers to cope with our mental health problems and every day this situation gets worse. Donahue and Nichtern[7] describe an experiment in training "teacher-moms"—mothers from the community trained to teach severely disturbed grade school children. Others have experimented in similar ways to recruit personnel. In the nursery teacher there is a rich reservoir of talent. Given a teacher with warmth and love of children and the capacity to be accepting and tolerant, add a facility for psychodynamic thinking with some knowledge of her own unconscious, and a nursery school teacher is properly equipped for training as a teacher-therapist.

On Involving Parents in Counseling

Our TNG parents are not so guilty or conflict-ridden about their children as are most parents who *bring* children to a child guidance clinic. And because of this, we seem to get *more* cooperation from them than has been my experience in other child guidance clinics. True, some of our parental education may be on a primitive level. We often have to suggest other ways to handle problems than the brutal punishment some of our TNG parents use. Many of our mothers have a concept of mothering that is limited to physical care alone; they have never learned the language of emotional care. Often simple suggestions about support or sympathy of a troubled child can be effective.

On Treating Preschool Children without Family Involvement

In several instances, without cooperation from the family and without a change in the home situation, we have effected a considerable change in the child. We appreciate the continually damaging effect a disturbed family can have on a child, but there are times when we can blunt this.

Marie was a pale, pretty four-year-old who spoke very little, grimaced a great deal, and seemed lost in a world of her own. She was unreal, inaccessible and phantom-like. She was being raised without a father or siblings by a schizophrenic mother who was unrelated and unresponsive. The mother was willing to have Marie in TNG but she herself wanted to have nothing to do with us. Our TNG teacher-therapist was given a specific therapeutic task—to make repeated spontaneous contacts with Marie. In several months her quality of relatedness to adults and children was drastically different. She began to make contact with other children and with her Day Care Center teacher. We had intervened and had stopped a process of autistic withdrawal. Same mother, same home, but treatment was effective to this degree. We work for family cooperation, but it is essential, with the nature of our families, that we have techniques that can sometimes be effective without that cooperation.

In summary, we have presented a method of treatment for preschool children which has the following assets:

(1) It seeks out children before they are brought for help as a preventive mental health measure.

(2) It treats a number of children simultaneously by the use of groups.

(3) It effectively treats children of underprivileged families by overcoming the reluctance of their families to seek help from mental health sources.

(4) It adds a new discipline to our forces by using the teacher-therapist.

(5) It can involve parents in counseling, yet it can sometimes be effective without active parental cooperation.

Our experience is in line with the growing trend toward a community mental health orientation. The clear implication is that when a community and its members do not seek needed help, the clinic must go into the community, for if we offer help early in childhood we may prevent the kind of behavior that fills our courts and fills our institutions and fills our headlines. We cannot allow such human waste.

REFERENCES

1. Hunt, R. G.: Socio-cultural factors in mental disorder. Behavioral Science 4:96-106, 1959.
2. Hollingshead, A. B., and Redlich, F. C.: Social Class and Mental Illness. New York, John Wily and Sons, Inc., 1959.
3. Freedman, L. F., and Hollingshead, A. B.: Neurosis and social class: Social interaction. Am. J. Psychiat. 113:769-778, 1957.

4. Redlich, F. C., Hollingshead, A. B., and Bellis, E.: Social class differences in attitudes toward psychiatry. Am. J. Orthopsychiat. 25:60-70,
5. Lambert, A., Mahler, M. S., and Moore, V. M.: A settlement house approach to community mental health. Public Health Reports 74:957-964, 1959.
6. Albert, A.: The treatment of emotionally disturbed children in a therapeutic nursery. Am. J. Orthopsychiat. 25:826-834, 1955.
7. Donahue, G. N. T., and Nichtern, S.: A school district for identification and adaptation to the needs of emotionally disturbed children. To be published.
8. Neubauer, P. B.: On the neglect of the young child (editorial). Am. J. Orthopsychiat. 33:777-778, 1963.
9. Bower, E. M.: Primary prevention of mental and emotional disorders: A conceptual framework and action possibilities. Am. J. Orthopsychiat. 33:832-848, 1963.
10. Slavson, S. R.: Analytic Group Psychotherapy with Children, Adolescents and Adults. New York, Columbia Universities Press, 1950.
11. Ginott, H. G.: Group Psychotherapy with Children: The Therapy and Practice of Play Therapy. New York, McGraw-Hill, 1961.
12. Albronda, H. F., Starkweather, J. A., and Dean, R. L.: Social class and psychotherapy. Arch. Gen. Psychiat. 10:276-283, 1964.
13. Westman, J. C.: Nursery school: Outpost for preventive psychiatry. Arch. Gen. Psychiat. 10:31-35, 1964.
14. Hunt, R. G.: Social class and mental illness. Am. J. Psychiat. 116:1065-1069, 1960.
15. Baum, E. O. and Felzer, S. B.: Activity in initial interviews with lower class patients. Arch. Gen. Psychiat. 10:345-353, 1964.

THE USE OF GROUP PSYCHOTHERAPY BY THE PRACTICING PSYCHOANALYST

By JOSEPH J. GELLER, M.D.

BOTH INDIVIDUAL psychoanalysis and analytical group psychotherapy have many features in common. From the standpoint of the patient, complete and uncensored communication, spontaneous presentation of thoughts and feelings, and the production of free associations are important in each. Similarly, the therapist must be honestly aware of his own feelings, he must make efforts to detect his distortions, and he must minimize the effects of his personal value judgments in both types of therapy.

In individual therapy, emphasis is placed on the ideas and concepts the patient presents. This content material is instrumental in arriving at the insight essential for successful therapy. In analytical group psychotherapy, more emphasis is placed on the processes going on within the group than is the case in individual psychoanalysis. Thus, in the therapy group although the content of the productions is important, considerable use is also made of process phenomena as an extra dimension of therapy. As in individual psychotherapy, free association is also utilized in the group situation. It serves to bring to awareness material which would otherwise remain at an unconscious level. In the group there is the further expansion of this material into a free play of processes, which brings into awareness covert and subtle feelings and emotions, and achieves a useful catharsis of some of the stronger emotional trends.

In both modalities, individual and group, awareness of the interaction between patient and therapist is important. In the group, this interaction, occurring in both directions, is of greater depth, extends over a broader range, and is more in evidence than in individual therapy. This leads to more productive therapeutic use of the material developed.

The group milieu thereby involves the patient in a broad, enriched, and active series of relationships which affords an opportunity for both qualitative and quantitative increases in useful data as compared with individual therapy. Along with these useful features are certain difficulties inherent in group psychotherapy. For the patient, there is a less intimate relationship with the therapist; resentments and rivalries may appear, due to the "sharing" of the therapist. The patient also has less opportunity to investigate his personal problems on a steady and intensive basis. Some extremely anxious patients may need to be handled on an individual basis until the acute phase of discomfort subsides.

For the therapist, there are difficulties too: he must be alert to a multitude of interacting relationships; he must make many more immediate decisions as to which of several courses of action to follow—or to avoid; he cannot readily stay abreast of the peripheral streams of activity; he will usually be drawn into a contagion of emotional reactivity somewhat in excess of the amount involved in individual therapy; and he is apt to be frustrated by more numerous lacks of closure.

Use of Group Psychotherapy in Private Practice

In private practice, group psychotherapy is most commonly conducted on a once weekly basis for a period of one and a half to two hours. Although in some cases it is used as the exclusive therapeutic modality, it is more often combined with individual therapy sessions. The number of these individual sessions ranges from several each week, in combination with the weekly group session, through one individual session per week, to occasional, intermittent individual sessions. It is the author's practice most commonly to offer one individual and one group session per week to those patients for whom group psychotherapy is being used.

Eight or nine patients have been found to be the optimum number for the group. Since psychodynamically mixed (i.e., heterogenous) groups are more useful than homogeneous ones, heterogeneity of diagnostic features and symptom presentation should be borne in mind when selecting patients for the group. It is a common fallacy about group psychotherapy that more effective results are achieved by putting psychodynamically similar patients into the same group; this does not seem to be the case in actual practice.

In selecting patients for a therapy group, the same criteria should be employed as those used to select patients for individual psychoanalysis. Motivation, ego strength, the nature of the defense system, the level of psychic compensation, and the potential for disabling anxiety are, for example, some of the things to be evaluated. In addition, however, it is specifically essential that the patient be willing and able to communicate with others.

There are also a number of specific indications for recommending a therapy group. Patients with problems of socialization can often use it to work out some of the anxieties involved in their isolation. Those who in their daily lives tend to get into difficulties because of the acting-out of defenses can use the group for this compulsive behavior. This develops both content and process material for therapeutic use and also helps modify the real-life acting-out, thus diminishing some of the patient's reality problems.

The patient who uses avoidance and denial as defense operations can also often be reached in the group setting. In individual therapy he does not produce the relevant material, even though the analyst may sense, and indicate to the patient, the probable existence of such mechanisms. In the group, interaction with other patients brings out those day-to-day occurrences which he does not "remember" to report in analysis. Occasionally a useful confrontation for the patient will also arise out of an illustrative episode developing within the group.

The group can act as a testing ground for the patient who is fearful of experiencing feelings. This applies both to areas involving hostility and aggression and to those involving closeness and warmth. By trying out the effects of letting feelings into awareness and by bringing them out toward the other group members, it becomes possible for the patient to look at the reasons for the suppression of these feelings. He will also tend toward some desensitization with improved reactions to the fearful emotions.

Group psychotherapy is also useful as an adjunct to definitive psychoanalysis. In spite of the intensity, thoroughness, and duration of a satisfactory course of analysis, adding a group therapy experience during the terminal phases can be of value. This is best done during the last year so that the diminishing analytic sessions and the group experience go on together. In this way not only does a broader spectrum of self-awareness develop in the patient but he also has an opportunity to test out some of his reactions in a setting where a high degree of consensual validation is available.

Use of Group Psychotherapy in the Community Agency

The psychoanalyst associated with a community agency finds his services used most effectively in the role of teacher, supervisor, and consultant rather than as an actual psychotherapist. If he is to fill this role as far as the group program of the agency is concerned, he must be familiar with both the theory and the practice of group psychotherapy: he should have a fairly thorough knowledge of the various types of group psychotherapy and their applications, as well as actual experience in their use. In many cases the psychoanalyst acting as therapy consultant to an agency has neither the interest nor the experience in group techniques to fill this role. It is then necessary for a second consultant to deal with the group program, a procedure which tends to destroy the uniformity of the agency's therapeutic approach.

Apart from its primary goal of meeting the patients' psychotherapeutic needs, the psychotherapy group greatly facilitates the staff training program of the clinic. The beginning therapist may be invited to join a group conducted by the consultant or some other experienced therapist. In this setting the beginner has a primary experience with a therapeutic situation, and can subsequently discuss his reactions with the supervisor. When the trainee is ready to conduct groups himself, the supervisor can sit in with him, and later review his observations and recommendations.

It may also be useful for the analyst-supervisor to conduct groups composed of staff members in which the participants become more aware of one another's basic value systems and communications symbols, and achieve a useful level of comfort and security with one another.

COMMUNITY MENTAL HEALTH

By GERARD CHRZANOWSKI, M.D.

Two Concepts of Community Mental Health

IN MANY RESPECTS our basic concept of mental health is structured after the classic Greek ideal of physical and mental harmony—an ideal difficult to practice. It is not easy to discover a formula of mental health broad enough to fit Barry Goldwater and Bertrand Russell, or encompass the requirements of a New York City taxi driver, a scientist and a minister. Love, happiness, relatedness, selfhood etc. are not concrete entities which everyone can acquire with professional assistance, and so become an idealized image of a sensitive, liberal, middle-class citizen.

Preventive psychiatry is used here as a concept which is relatively free from a preoccupation with the assumed ingredients of mental health. The emphasis is rather on the patient's individual life situation and his functional integration with the group to which he belongs. One of the purposes of preventive psychiatry is to study repetitive stresses, misunderstandings and miscommunication before the distortions reach pathological proportions.

Reevaluation of Therapeutic Methods

Many of our therapeutic procedures and goals are not geared to the needs of large numbers of patients. There has been a shift from dyadic to group or mass phenomena and to social goals related to community mental health and community psychiatry. This calls for a genuine effort to apply thoughtfully some of the findings gained from psychoanalytic research and practice on a larger scale.

Community mental health cannot be successfully dealt with by merely training more psychiatrists and auxiliary specialists and by opening a large number of better and bigger treatment centers. Beyond utilizing all available community resources we must also re-examine some of our basic technical and theoretical foundations. Efforts must be made to utilize potential (environmental factors intsead of working primarily on symptoms and end states. There is much wasted effort involved when we have to return people to the same situation which contributed largely to their difficulties, since it is doubtful that psychotherapy can expect to immunize people against recurrent exposure to stressful situations. Different socioeconomic standards also call for different methods in coping with stress, and it is doubtful that one type of psychotherapy can cover every cultural level.

Comparisons between Psychoanalysis and Community Mental Health

In psychoanalysis, the legitimate needs of the individual patient are accorded primacy over communal demands whether they be made by the state, the family or any other organized form of power. Self-expression and individual freedom are explored against the background of formative emotional experiences and social controls. However, income level, education and verbal facility rather than the psychiatric diagnosis determine the choice of analytic treatment (Hollingshead and Redlich): i.e., analysts in private practice work with patients who are similar to themselves in significant respects. Yet the therapeutic process is always a form of social control operating through a socio-relational system (Talcott Parsons). An individual's capacity to live with and among other people comes in part from the appreciation that a person and his society have reciprocal expectations of each other. There is an equilibrium, or homeostasis, in every social encounter which links the participants with each other in a transactional interplay. Community mental health attempts to cope with the individual in a broader context than does psychoanalysis. The inner life of the patient, his resistances and defenses are more clearly viewed in the light of situational determinants. External factors related to the patient's family, social and community life are given special consideration.

Many patients in a community clinic have been exposed to life situations which differ significantly from the life-pattern of the therapist. There is a much greater tendency among clinic patients to assume the familiar role of the typical medical patient. The therapist is often placed in the position of an undisputed authority who is expected to tell the patient what to do. More often than not, the patient will assume a submissive attitude, which hinders genuine participation on his part.

To see this formidable barrier as a primary function of the patient is unfortunate. The maintenance of social distance is to some degree mutual.

There is no doubt that a Community Mental Health Clinic must cope with a distinctly higher percentage of severe mental disorders than does a private analyst in his practice. Many patients have great difficulty in keeping a job and in providing adequate support for their families, and it becomes urgent to help them regain social and economic competence as soon as possible. A Community Mental Health Clinic must provide treatment for the large number of people who require psychiatric attention and do not receive it, either because of ignorance, personal resistance, economic factors or because of a lack of other community facilities. The Midtown Study by Rennie, Opler et al., indicated that the lower social groups, which receive the least psychiatric treatment, had the highest incidence of serious mental disorders.

Experience in a Community Mental Health Clinic

During the last fifteen years in a Mental Health Clinic I have conducted close to 3,000 psychiatric consultations, made spot checks on patients whom I had seen initially and have re-evaluated my diagnosis and prognosis. Our Clinic population consisted almost exclusively of a low to very low income group, blue collar workers and others in marginal economic positions, with few lower middle class representatives. In order of frequency the complaints were: marital problems, disturbed relationships with children, maladjustments at work and, in some instances, impulsive and uncontrollable behavior. We found a very small number with overt sexual problems, but there were some alcoholics, drug addicts and homosexuals, and a fairly large number of borderline psychotic patients as well as some with overt schizophrenic conditions. The children referred to us had difficulties at school or exhibited antisocial behavior. The greatest influx occurred shortly after the parents had received report cards, which constituted a warning that impelled parents to action after long neglect.

We were able to confirm the findings of Whitehorn and others that certain therapists consistently had better treatment results than others; i.e., a high dropout rate in a Clinic is not a reliable index of the patient's motivations but of the quality of the therapeutic staff. In this connection, I found that therapists whose expectations were relatively objective and who were able to set realistic treatment goals on the whole had best results.

It is a requirement for the staff of this particular Clinic that every therapist undergo some form of intensive personal psychotherapy or psychoanalysis, as still by far the best preparation for the skillful therapeutic use of one's personality. Since authority can also be used constructively in a clinic milieu, we arranged consultations with the Director of the Clinic for many patients who encountered difficulty in treatment, with an effort made to have the therapist present; personal attention thus accorded to the patient not infrequently provided an opportunity for illustrating certain situational difficulties in a potentially problem-solving frame of reference.

Some patients tend to become habitués and integrate their attendance at the Clinic into their life situation in a highly dependent fashion. The majority of patients who failed to show a tangible improvement in one and a half years made little progress thereafter. Most patients in a low cost clinic, asked why they terminated treatment, replied that it was primarily because of a lack of money.

Technical Aspects and Implementation in a Community Program

Interviews with the patient should be reasonably well-structured and should have, if possible, an educational frame of reference.

> I introduce myself to the patient and assure him that his earlier social service intake interview serves me as a frame of reference. My opening remark will usually be along the line: "Mr. X, one of our social workers tells me that you are here because of certain complaints. Did I understand her correctly? Could you tell me more about the situation? And may I hear whether she left out something important?" Then I usually go on to inquire as to what goes relatively well for the patient, what he considers to be potentially rewarding and pleasing and what causes difficulties. It often winds up with, "If all goes well, what do you hope or expect to get out of coming here?"

The procedures are geared to the patient's style of life and are designed to help us understand the person's particular needs from his point of view. Generally, our middle-class groups have prestige problems, tend to intellectualize and have difficulties in terms of role commitment. In contrast the less privileged groups are much more threatened in elementary aspects of security. They require a different and more reality-oriented kind of reassurance, and they also require a particular kind of education. Carefully selected films and pictorial illustrations can be of considerable merit, particularly in view of the widespread distortions of our media of mass communication. There is much emphasis on unrealistic aspects of human relations and a great deal of misinformation concerning common causes of stress and strain. Furthermore, the introduction of group orientation programs and demonstration set-ups can be helpful. There is frequently merit in having as many family members present as possible, or in conducting the interview in a small group. Many forms of group therapy and family-oriented therapy have proved to be most helpful in a clinic setting. In some instances tangible results can be observed in a remarkably short period of time.

Body reorientation programs (Gindler and Jacoby) and various forms of exercises may be used as particularly applicable to a group of people who are action oriented and to some degree distrust words. Such a program can be offered on an individual basis as well as in groups. Finally, there may be a place for an evangelistic kind of approach, which has been so helpful in Alcoholics Anonymous. I also favor the use of people, who are somewhat similar to the patient in background and who work under the supervision of clinically experienced specialists, as therapeutic agents. Chemotherapy may also be judiciously integrated into the overall clinical program.

A word about the rapidly increasing number of comprehensive Mental Health Centers. They have become the glorified model of an integrated, all inclusive approach to the problem of mental illness. There is no question about the contribution which such a giant set-up can offer in providing almost unlimited service programs as well as superior opportunities for research, teaching and rehabilitation. However, centralization tends to foster bureaucracy and automation, and may lead to the creation of psychiatric supermarkets and mental health department stores. There is little likelihood of doing justice to strictly personal, individual human difficulties in a super-efficient but socially and humanly artificial atmosphere. The sociohuman climate of a highly organized center tends to a stop-gap, symptomatic approach without sufficient emphasis on preventive aspects. It also tends to reinforce the predominantly medical role of the patient, who does not think of himself as an active participant in the therapeutic endeavor. Such centers should therefore not have complete autonomy and it seems best to combine them with other mental health facilities in the community.

THE GOALS OF COMMUNITY MENTAL HEALTH CONSULTATION

By ARLENE WOLBERG, M.S.S. AND
ELENA PADILLA LAWSON, PH.D.

PSYCHOANALYTIC BACKGROUNDS OF THE CONCEPT OF MENTAL HEALTH CONSULTATION

IN 1924, Freud wrote about modified psychoanalysis and consultation services as patterns of future care derived from psychoanalytic technique to serve the large masses of people. He stated:

> Now let us assume that by some kind of organization we were able to increase our numbers to an extent sufficient for treating large masses of people. Then, on the other hand, one may reasonably expect that at some time or other the conscience of the community will awake and admonish it that the poor man has just as much right to help for his mind as he now has to the surgeon's means of saving life; and that the neuroses menace the health of a people no less than tuberculosis, and can be left as little as the latter to the feeble handling of individuals. Then clinics and consultation departments will be built, to which analytically trained physicians will be appointed, so that the men who would otherwise give way to drink, the women who have nearly succumbed under their burden of privations, the children for whom there is no choice but running wild or neurosis, may be made by analysis able to resist and able to do something in the world. This treatment will be free. It may be a long time before the State regards this as an urgent duty. Present conditions may delay its arrival even longer; probably these institutions will first be started by private beneficence; sometime or other, however, it must come.
>
> The task will then arise for us to adapt our technique to the new conditions. I have no doubt that the validity of our psychological assumptions will impress the uneducated too, but we shall need to find the simplest and most natural expression for our theoretical doctrines. We shall probably discover that the poor are even less ready to part with their neuroses than the rich, because the hard life that awaits them when they recover has no attraction, and illness in them gives more claim to the help of others. Possibly we may often only be able to achieve something if we combine aid for the mind with some material support. It is very probable, too, that the application of our therapy to numbers will compel us to alloy the pure gold of analysis plentifully with the copper of direct suggestion; and even hypnotic influence might find a place

in it again, as it has in the treatment of war neuroses. But whatever form this psychotherapy for the people may take, whatever the elements out of which it is compounded, its most effective and most important ingredients will assuredly remain those borrowed from strict psychoanalysis which serves no ulterior purpose.

This concept had notably more direct influence on the formation of modern mental hygiene clinics than on the mental health consultation modalities as they are known today; however, it was the germinal notion of consultation in psychoanalysis. One of the main avenues for the development of community mental health consultation emerged from the increasing role in community service and the involvement of psychoanalysts in the mental health movement. Other trends also affected the direction of community mental health consultation.

Awareness of the potentials of mental health consultation undoubtedly received impetus from the consultation service (MHCS) of the armed forces. More recently, consultative services to local health and welfare agencies, schools, and law enforcement agencies have been listed among the basic components of a mental health service by the Joint Commission on Mental Illness and Health (1961) and the Ad Hoc Committee of the Surgeon-General (1963).

Psychoanalysts have spent many years in becoming proficient in the clinical field. Their training goes a long way in bringing understanding to bear on individual dynamics and they are expert in handling patients in the treatment room. But often the community seeks psychoanalysts out for another type of service—that of helping both professionals and nonprofessionals in dealing with emotional problems as they are expressed in nonclinical situations. Psychoanalysts in these circumstances are at a disadvantage with their particular orientation, for they usually do not practice the kind of skills that are necessary to help others who work with people who may have emotional problems but who are not seeking to become patients. The educational skills of the average psychoanalyst tend to be organized around the clinical model, and this is both confusing and inappropriate for nonclinical professionals. The quandary is further intensified when it is realized that the basic problem with personnel who seek consultative services is the lack of understanding of mental health concepts and information when using mental health techniques relevant to their own work. The clinical model has proved to be unsuccessful as a basis for consultation work.

Toward an Operational Definition of Community Mental Health Consultation

Since mental health consultation has roots in the psychoanalytic movement, consultation has had a strong "clinical" orientation with emphasis on evaluation of individual cases. Other trends which have influenced the activities of community mental health consultation have led in other directions, such as organizational, managerial and industrial consultation. The current concepts of consultant and consultant's role are used with diverse and even contradictory meaning and reference. This has led Haylett and Rapoport (1964) to describe both the title and role of the consultant as ambiguous within the mental health field itself. This ambiguity of role and title can be reduced, if not eliminated, by developing an operational definition of the concept which permits clarification of the domain of both consultation and its technique. Comparisons and contrasts between psychotherapy and consultation can then be more readily identified. At the Postgraduate Center for Mental Health, community mental health consultation is defined as essentially an educational process involving a temporary relationship between a consultant and a *functioning social unit* which may be called a *help-needing-system* or a *client-system* as distinct from an individual or personality system. Lippitt makes this distinction, and an adaptation of his model can be used in the mental health field.

The process of mental health consultation therefore is viewed as bringing particular skills and information which have to do with the field of mental health to bear on problems arising within a client-system which interfere with goal fulfillment, role performance, and the personal satisfaction of the members of the staff; the goal is to identify problems in these contexts and to find appropriate solutions for the *client-system,* e.g., the staff and the administration.

The methods of consultation are organized around the processes of entry into, attachment to, and withdrawal from client-systems. It is in this context that the particular resources and skills of psychoanalysts are utilized in community mental health consultation. This leads to the use of specific techniques geared toward:

(a) Assisting the client to discover the nature of the problems, to make a diagnosis, and to help clients set up a problem-solving plan, thus facilitating and enhancing the problem-solving skills of the groups participating in the client-system;

(b) Opening communication channels within the client-system and handling the effects of feed-back, resistance to change and to learning new ways of coping;

(c) Helping to implement the plan for solving problems within the client-system;

(d) Helping the client-system to obtain the appropriate extramural or "outside" resources when these are indicated.

From this working definition and body of activity it appears that theories of clinical practice are not wholly adequate for the tasks faced by psychoanalysts and other therapists in consultative roles. This definition, therefore, anticipates a practice theory for community mental health consultation on which to base both a special training program for psychoanalytic mental health consultants and a mental health consultation service to the community. Training must take into account the goals of consultation. There are several kinds of objectives which can be met, some more effectively than others. The following are three different kinds of projects illustrating three approaches. The first is perhaps the most efficient from the point of view of the organization, or the client-system; the second gives information about a problem and promotes some changes in the attitudes of the personnel in the client-system about the problem, but it does not leave them with techniques for day to day working; the third is least helpful in that it merely gives a diagnostic impression, but does nothing to help solve the problem.

A. Community Mental Health Consultation Program for Staff of a School Unit Working with Absentees

This project was undertaken by the consulting supervisory staff of the Postgraduate Center as a public service. It was also an opportunity for consultant trainees to observe the total experience of consultation from entry thru conclusion. The consultation service of a community mental health center can touch the lives of literally hundreds of people who might otherwise not be reached or exposed to mental health service. The following is a case in point:

The Client-System

The casework staff of the case study unit of one of the Bureaus of the school system approached the Center. Their main concern is regular school attendance by pupils, particularly the hard-to-reach children with too much anxiety to maintain the one-to-one casework relationship. In 1960-61, about 195,000 children were served by this Bureau. The social work staff was involved in home visits, referrals to community agencies, close supervision of children, consultation with community agencies and appearances in court. This particular staff felt that they needed new techniques to deal more effectively with the problems of the "hard to reach" children so that they would increase their rate of attendance, improve their school work and continue their education. The Bureau had developed a classification of the personality characteristics presented by both the parents and absentee children which suggested that the parents in any given instance could be overprotective or rigid and punitive; others were apathetic, ineffectual and/or inept, or their attitudes toward education were negative or they were emotionally disturbed or overtly psychotic. The children showed anxiety reactions to parental attitudes and/or marital discord, had evidenced tension from school pressures (peers, teachers, curriculum), used absence as a weapon against parents, engaged in "acting-out" behavior with members of peer groups (gang, etc.), and some were also frankly psychotic. It was on the basis of their own recognition of the emotional maladjustments underlying school absenteeism that the casework staff of the Bureau approached the Postgraduate Center for Mental Health.

Entry Into the Client-System

This began in December 1960 when the Bureau initiated its request to the Center. After a series of conferences it was agreed to organize a training program consisting of four broad phases:

Phase 1: This would consist of six to eight exploratory sessions at the Postgraduate Center during which the case study supervisors of the Bureau and consulting supervisory staff of the Center would meet to discuss typical cases handled by the Bureau, the kinds of children the social workers felt would gain most from group experience, the evaluation of the broad problems involved in setting up a group counseling program within their particular work setting, and how the consultant would go about providing an initial orientation in the use of the group counseling method within casework processes. It was anticipated that at the end of this beginning phase, a set of criteria for the choice of clients to participate in groups and a method for handling practical problems would be discussed and described.

Phase 2: This would include a regular fifteen-session seminar for the case study supervisory staff of the Bureau and those caseworkers who were interested in taking the formal training in group casework and who would handle the first trial casework groups. The focus here was to integrate concepts from group dynamics and casework, and to discuss how these could apply in the specific group work situation with children. This phase of the training was a regular seminar course at the Postgraduate Center and was accredited by the Board of Education.

Phase 3: This aspect was planned to encompass the actual operation of the first trial casework group. Training in this phase would be directed primarily toward the Bureau's supervisors who would, in turn, supervise the professional staff of the group casework program. The Postgraduate Center organized "consultation groups," each composed of a consultant from the Center, a supervisor from the Bureau, and one of the social workers for whom the supervisor was responsible. Consultant trainees from the Postgraduate Center, and supervisors and workers from the Bureau could participate as observers and could also join in the discussions. The role of the consultant was twofold: to teach the worker how to conduct the group and to teach the supervisor how to supervise the social worker carrying the group. It was agreed that beyond the formal training program there would be a need for continuing consultation between the Bureau supervisory staff and the consulting supervisory professional staff at the Postgraduate Center. This phase was to continue for a period of two or three years in a progressively diminishing manner as the Bureau staff gained greater skills and self-confidence in the use of the method.

Phase 4: It was also estimated that after four or five years the program might be expanded to train social workers in this Bureau for assuming responsibility for a city wide group program in the schools. The consultant would continue as in Phase 3 to work primarily with supervisors. At the fulfillment of this goal, the Postgraduate Center would withdraw from the client-system and the consultation program would be terminated.

A consultation method was worked out to accomplish this goal. The plan of the project called for each caseworker to see two groups of children each week over a period of one semester (fifteen weeks) in addition to his regular work; each group was to consist of eight children. It was assumed that in the future, twenty caseworkers would see sixteen children a week in addition to their regular caseload, that this would mean that their capability would be expanded to help 320 additional children in a semester. The mental health consultation was not to try to make group therapists of these social workers, but was to attempt to adapt group methods to casework procedure so that these workers could, at the same time that they brought the absentees back to the school system, do something active to enhance the mental health of these children. The consultant role would include training the caseworkers to use techniques which could be adapted to the needs of the Bureau, particularly with the "hard-to-reach" client or the chronic nonattendant. Group methods appeared to be more successful than the one-to-one relationship with this type of child although children were seen for one or two individual sessions if they so requested. Before the first trial group meeting of children was started, it was assumed that younger children ten to twelve years old would provide an optimum learning experience for the first group casework to be held, and after the caseworker had gained some experience he then could try an adolescent group. The initial criteria used in selecting children were: equal number of each sex; eight children in a group; a minimum I.Q. of eighty; ages ten to twelve; absence for at least twenty-five times in the preceding year; a history of absenteeism for several years; a balance between acting out children and withdrawn children. The decision was made to exclude psychotic children.

One of the directors of the Bureau questioned whether there might not be the possibility of re-enforcing the disturbances of the children in the group since they could, in effect, secure additional support for their acting-out behavior from other group members. It was conceded that this was a possibility and that ways and means of meeting this problem would be a major consideration in the training program. Insofar as the first group was concerned, this problem would probably be minimized since the social worker would only be working with younger age children.

It was clearly stated that casework group sessions would not be aimed at conducting clinical groups as in psychotherapeutic sessions even though it was recognized that the children had emotional problems; rather, the program would revolve about discussion of problems related to school attendance and also about other difficulties which the group members might be willing to consider. Basically, the program would be focused at improving the general understanding of the members of the group, and in having them discuss the possibility that school absence has a relationship to personal problems—worries or troubles of one kind or another. A casework approach would be maintained throughout, the caseworker would not discuss or disclose clinical interpretations and motivational dynamics with the children's group, although such interpretations were to be made to the caseworkers in their own seminars at the Postgraduate Center and in the consultative relation so that they would understand the latent aspects of the children's school attendance behavior.

Attachment to Client-System

In October 1961, nine months after the initial contacts with which entry process started, the first seminar for caseworkers and their supervisors was started at the Postgraduate Center. By this time, also, the children's groups were organized. Three separate beginning classes have already been given each of the past three years, and consultation sessions have been ongoing concurrently. The classes consist of a fifteen session seminar; members of the seminar are organized into small discussion groups in which materials from the children's groups are used. As a result of the learning experience, the basis for selecting children for the group casework sessions have been broadened to recognize that older adolescents as well as younger children with serious emotional difficulties can benefit from this type of group experience. This consultation project is still in this stage since the consultation goals have not yet been achieved.

The goals thus may be restated: to train a core of personnel in the Bureau to be expert in the group casework methodology and to train supervisors who could, in turn, take the responsibility for training future caseworkers in the method as it applies to school absentees.

The overall supervisor of caseworkers in reviewing children's groups made the following empirical evaluation of the Group Casework Program:

School attendance: Caseworkers felt that on the whole there was marked improvement in actual school attendance by participant children. Also observed was an awareness on the part of the children that unjustifiable absence for whatever reason was not a desirable trait. The importance of this observation is that this was a new value set by themselves rather than by the adult world.

Scholastic achievement: There was at least an average change anticipated for the grade; the students have been able to keep up with their grades even though they have had more absences from school than the other children.

Attitudes toward school: There appeared to be some growth in depth for the adolescents, but not in the younger children who, on the other hand, showed more visible change as measured by commendations received for school citizenship, etc.

General attitudes to the world, to adults, and to peers: All areas of change were positive. The preadolescent group particularly showed change in the ability to communicate freely with each other in the group as well as outside the group. Of the two adolescent groups, it is anticipated that all children will be graduated from Junior High School, a condition that might not have prevailed but for the group casework program.

Attitudes of school personnel: The group caseworkers have reported that in all cases, school personnel have been enthusiastic about the outcome of the groups in terms of the factors discussed above. In all cases, also, the principals wished for a continuation of the program.

In one group, the caseworker asked the children to evaluate the group. These children felt that the program was helpful to them because the caseworkers were not critical of them, because they were helped to understand themselves, and because it helped them get away from childishness. It is noteworthy that the children in this group all recommended that this kind of a program be given to all school children. Some of them also indicated that they would be ready to participate in such a program after school hours.

As stated, the consultation goal for this program was the progressive training of supervisors and caseworkers in group casework methods so that these techniques could be made a part of the skills of the social workers and attendance teachers in the client-system.

B. Community Mental Health Consultation Program for Schools with Special Classes for Disruptive Children

At the request of the head of a program of special classes for children considered disruptive in regular classes, a consultation project was initiated by the Postgraduate Center. Prior to the inception of the project, a number of conferences were held between some of the Center's supervisory staff and members of the staff of the school system in the central office, and in one district office. A study was made of school disruptive behavior and the characteristics of the children with whom the teachers had most difficulty.

Specifically, the project was addressed to three grammar schools in a Manhattan school district. These schools operated special classes for nine- to eleven-year-old children classified as "disruptive."

It was decided that the MPAG, a multilevel-planning-activity group technique developed at the Center for creating, improving and facilitating communication among component groups of a client-system, would be used in this project. It appeared from the entry process and preliminary assessment of the problem that, in this instance, the critical problem and target of consultation was the dearth of any rapprochement among children, teachers and administrators.

To achieve this objective, a plan was developed to organize a children's discussion and planning group program, a teachers' workshop program, and both informal and formal discussions with the administration. Through a children's plan for school improvement, the children, teachers and administration would have *one common area of involvement* around which an educational and emotional growth experience could be effected. Three community mental health consultant trainees under supervision were assigned by the Postgraduate Center to this task. Their consultative functions included efforts toward improving communication and understanding among the three component groups in the school, *the students, the teachers* and *the administrators;* this goal was part of the MPAG concept: the *multi-level aspect* of the consultation process.

The three schools chosen were under the management of one assistant superintendent. The district is part of a large area of Manhattan which had been undergoing rapid social and physical changes. Population movements, both in and out of the neighborhoods, had been transforming the character of the neighborhood. The urban renewal process had already made its imprint with new housing projects and rehabilitation of old buildings. Simultaneously, slum tenement formation had been taking place in many sectors. Large apartment houses had been transformed into rooming-houses for poor families and unattached persons and, in relatively few months at a time, whole blocks had been turned into slum housing. The district consisted of sets of decaying and rising neighborhoods with little cohesion and social integration. Leadership provided by the school, professionals, civic, political and religious organizations, although active, had not achieved adequate grassroots effects. Schools in the district had a high proportion of children considered predelinquent and disruptive by the school authorities.

The kind of community organization was reflected in the school in the type of students and their backgrounds, in the teachers' feelings toward the children and in the attitudes of the principals that educational goals must be limited.

The schools were far from uniform. The particular kind of authority system and the patterns of relations set by the administrators influenced the content of problems expressed in the groups of teachers and children; they also influenced the course of the consultation process and the role of the mental health consultant. The unresolved problems of the trainee-consultants with respect to authority were evident in many situations, and these were to affect the course of the projects. Working within the framework of MPAG, the consultant-trainees had to introduce certain modifications to suit the particular circumstances of each of the schools.

The program was planned for one school semester.

One children's group of six to eight children was organized in each school with the aim of first encouraging discussion of how they could improve their school; second, to give them the opportunity to ventilate their feelings about school, teachers, principals and parents; and third, to agree upon a plan for school improvement to be discussed with their teachers and principal. By agreement among the teachers, principals and the children, the latter could carry out the plan within the school system. It was felt that if these children could be part of a group which could follow through on implementing a constructive plan, and the school authorities and teachers would become aware of it, that the experience would stimulate a kind of learning that would enable them to change some of the disruptive attitudes they had toward the teachers, the administrators and toward each other. A workshop or seminar was organized for the teachers so that selected material from the children's group could be discussed in the teachers' group, so that knowledge thus obtained would enable the teachers to improve their classroom techniques and ability to deal with these children. Teachers received increment credit for this course which centered upon the dynamics of disruptive behavior in school, the role of the teacher, and techniques for coping with disruptive children.

With the help of principals and teachers, candidates were selected for the children's groups in each school. The teachers rated children on the rating sheets which had been developed from the study of teachers' descriptions of children's disruptive behavior as reported in school records. Each child in a roster of disruptive children was interviewed by the consultant. The consultant finally decided which children would be chosen on the basis of their interest in participating in the groups. Teachers and principals were informed of the final selection of children in writing. Ten weekly meetings were planned with the children. Through the principals in each school, teachers were invited to participate in the teachers' workshops. In school A, the principal decided which teachers would be permitted to participate, as some of the teachers of children in the group had other assignments. In schools B and C, the mental health consultant was permitted to choose the teachers with the approval of the principals. It was assumed that it would be better that the groups should meet without the influence of the so-called authority figures being felt in the group meeting. This plan seemed to have been accepted, but after the first group session, the principal of school B dropped into the teachers' meeting and took over the group discussion, and the assistant principal of school A also attended a meeting. In both cases it was necessary for the consultant to deal with the anxieties of the administrators and to reassure them that they would be apprised of all that went on. It was decided that their ongoing participation would be effective only if more regular contacts were maintained by the consultants with the administrators. The principal of school C showed his concern to the consultant by indicating that the children became very upset after the "permissive situation" of the group discussion, and he felt the need to come in to see that the children behave properly when leaving the group to return to their classes. In this case, the consultant discussed with the principal the possibility of the group preparing the children for their transition into the "normal" school situation. The anxieties of the consultants were aroused in these situations, too, and often they developed negative feelings toward principals, and had untoward emotional reactions which had to be discussed and analyzed with their supervisors.

The themes of teachers' workshops in the three schools gravitated to the teachers' interest in issues such as how to maintain discipline in the classroom; the role of the teacher (how can she teach disturbed children? how can she be a "nursemaid, psychiatrist and teacher" all in one?); how to cope with feelings of helplessness in efforts to deal both with the school administration and with disruptive children; how to develop methods of helping children gain mastery in a variety of subjects; and how to bring themselves to love and care for difficult children who in any case are soon to go to other teachers. In the children's groups, the themes were: how can we communicate with our teachers? how can we obtain recognition from the teachers and from the school administration? how can we get along with other children? how can we get along with our parents? The theme of punishment also appeared recurrently and the children expressed the wish to move about in the classroom and to do so freely with the approval of adults.

The group situation was anxiety provoking to the younger children. In one group anxieties were alleviated when the children got up and walked around the room to play a game. Some children would leave the room when they became anxious, but they returned after a short time. Testing of the mental health consultant's authority and his capacity to exercise it, and provocation to punitive action was also a recurring pattern of behavior. Physical restlessness seemed to be used to secure additional attention. Two of the three consultants from schools B and C had to move from a "permissive" role to an active authority-exercising-role, restraining children who were disruptive. In the group in school B, one child told the consultant who had been acting "permissively" that "the time has come when grown-ups have to do something."

The idea of inviting their teachers and the principals to term-end parties was the way in which these younger children conceived of constructive action, by communicating with their elders who were in authority positions; this idea was discussed in all groups and welcomed by the teachers. In school C, the principal frowned upon the idea of attending a party "at the children's request" and called the assistant superintendent to inquire about the wisdom of such a plan. The assistant superintendent commented that the children were expressing a wish to be closer to the teachers and the principal, and after this interpretation the principal went along with the children's request.

In assessing the effect of the consultation project on children, teachers were asked for comments; they reported changes in the general appearance and grooming of children participating. Specific children were said to have improved their behavior and to have become more interested in their studies; such changes were noted to have appeared after the children had developed a positive relationship with the mental health consultant. Children themselves said that they had been feeling better about school, that they liked school better and that they had learned to like their teachers. Principals, other officials, and the assistant superintendent were pleased with the way the groups had helped teachers and children express their desire to get to know each other better; they also hoped that these relationships would continue. Constructive suggestions also came from school sources requesting that the children's groups be conducted during the last school hour, as they felt children could cope more adequately with anxiety and restlessness stirred up by group discussion if they did not return to the classroom; this would also simplify the problem for the teachers who would not have to deal with the activity of the children who returned from a "permissive environment" to the formal classroom where more authoritative measures were utilized.

In many respects this project was successful, but yet it did not accomplish the goal of involving the school personnel to give them actual techniques for helping disruptive children and to make such techniques part of their teaching and supervisory skills. The client-system welcomed the program in a passive way, as a service received and from which gains could only be obtained if such a service was given. In part, the complexities of the school system and the shortness of the program may account for limiting the consultation goals as noted above.

C. Community Mental Health Consultation to a Social Agency Responsible for a Day Program and Foster Home Supervisory Program

This project involved a small social agency dedicated to the care of foster children. The agency was divided into two sections: a boarding school in the country for 100 children between the ages of one and seventeen, and a department located in the city supervising 150 children placed in foster homes. The staff consisted of nine persons: one supervisor and three caseworkers, the administrator, two clerks and two assistants. The total administrative budget for the foster home department amounted to $18,500 a year. The foster mothers belonged to working class families, and were, for the most part, in middle age; a few had finished high school, the others had less education.

The request was initiated by the foster home department. The director was interested in having the caseworkers develop skills in handling problems with the foster mothers; for example, how to give information which would serve as guidance for foster mothers educating adolescents in sexual matters; how to clarify the educational policies of the agency to the foster mothers; what techniques to use in the handling of the anxieties of the foster mothers in relation to special problems, such as how late the children should be allowed to stay out at night, or what to do about antisocial behavior.

The method used by the community mental health consultant was to hold a series of meetings with the staff so that they could give him their impressions and suggestions as to possible solutions to the specific problems for which the consultation was requested; he studied several sets of records in order to familiarize himself with typical case histories and to learn about the administrative steps and the procedures used in bringing a child from the community to the care of the foster home placement agency; he became acquainted with the methods of the Department of Welfare, the Children's Shelter and other organizations, so that he could understand the meaning of experiences with these organizations to the foster child; he learned the criteria used for the selection of foster homes; he had a series of discussions with a group of foster mothers, after which he met with the administrator and the supervisor prior to having a series of conferences with the social work staff; and finally, he met three times with the social workers to discuss the way in which they were handling problems with the foster mothers.

The problems raised by the mothers involved: ways of winning a child's confidence; problems concerning the giving and the receiving of affection; ways of giving recognition and status in the family; problems regarding sexual promiscuity—how to prevent promiscuity, how to teach appropriate role relations with the opposite sex, how to reduce fears in adolescents who have been sexually traumatized (raped or molested by adults); how to handle antisocial behavior such as boredom, insolence toward teachers, fights in school with other children, tantrums, enuresis, stealing, and in one case a problem of pyromania; what to do about family problems related to divided loyalty of children for foster parents and their own parents; how to talk about the absentee foster father; what to do about money allowances; what to do when a child had a reading disability; what to do when a child needs psychotherapy.

The consultant felt that further training was indicated for the social workers, who were not sufficiently familiar with community resources, had hopeless feelings about their clients which they projected onto the parents and the children, were deficient in the clinical understanding of the emotional significance of enuresis, pathological lying, fire setting etc. They also had unresolved problems with authority which interfered with their relationships with the foster mothers and thus in their ability to interpret the goals of the agency. The consultant found a generally high cultural level in the foster mothers, and a great deal of understanding for children. In their contacts with the consultant, the foster mothers were strained at first. They came to the meetings uninformed as to their purpose. They expected to hear some sort of educational talk stressing admonitions. At first they were not willing to verbalize their problems since they felt this was the equivalent of admitting in the presence of a representative of the agency that they did not know how to handle the child; they wanted recommendations from the "expert" and did not want to listen to each other. After the first few meetings, however, they were able to recognize the value of sharing experiences.

The consultant wrote a report recommending advanced casework training for the social workers and he suggested that the agency obtain the services of a psychotherapist who specializes in adolescence, asking him for a course stressing mental health information about the various problems of adolescence and group skills which could be adapted to the casework process.

To summarize: Three different community mental health consultation projects undertaken by the Postgraduate Center on behalf of agencies with responsibility for children have been presented to illustrate issues related to the goals of mental health consultation. All three programs required joint study and assessment of problems; two involved joint planning of programs for changing the existing situation; and one included an organized learning activity for staff and supervisors including practice in the use of special techniques with their own clients. The specific goals of each of these consultation programs varied with the kind of planning and agreements which could be worked out, and the kinds of involvement and commitments of the clients for introducing and supporting changes in the client-system. To be noted is that the evaluation of these programs is based on the opinion of participants and on professional judgment. The ultimate test of the effectiveness of community mental health consultation as well as that of all mental health services, however, lies in the development of adequate scientific evaluation, and one of the prerequisites of such evaluation is the establishment of rigorous programs in which the steps and activities are explicitly identified and described.

Training Program for Community Mental Health Consultation

In addition to treatment, teaching and research, community service has been one of the functions of psychoanalysts, particularly of those in mental health clinics, in teaching hospitals and in professional schools. The magnitude of community problems and the responsibilities that they demand are beyond the scope of traditional psychoanalytic training. In view of this, the Postgraduate Center has evolved a specialty training program in community mental health consultation.

In the present stage of knowledge, community mental health consultation practice theory draws from the behavioral and biological sciences as well as from the practice theories of administration, medicine, psychiatry, psychoanalysis, clinical and industrial psychology, public health and social work; thus the formal training of mental health consultants requires a variety of multidisciplinary approaches. This can be made viable only by a multidisciplinary faculty drawn from the above fields. The emphasis of the training program is on areas which call for special professional and technical competence in the consultative roles. For example, sociological concepts and methods for the study of social systems are necessary for the community mental health consultant working on the problems of a service agency or facility. Communication and organizational skills in teaching, in group methods, in supervisory techniques, and in clear presentation and writing of reports and educational materials are also necessary. Furthermore, the community mental health consultant has to be familiar with prevailing social problems in the community and with their implications for mental health, i.e., delinquency, alcoholism, school problems, drug addiction, marital and family problems, suicide, homosexuality, unemployment, chronic diseases, poverty, etc.

In delineating the theory and technique of consultation, it is necessary to draw from many diverse theoretical and empirical works. Sociometry, group therapy and role techniques are of particular usefulness since group methods and social interaction processes are essential to the technique of consultation. The contributions of Bertha Reynolds (1942), Jessie Taft (1940), Charlotte Towle (1950), Virginia Robinson (1930) and Lucille Austin (1950) in social work are required reading. Aptekar's (1955) work is used as a reference for understanding Rankian theory. The theoretical bases of empirical research in other areas have been developed primarily from the works of Merton (1947), Parsons (1937), Homans (1950), Cartwright and Zander (1953), Bales and Strodtbech (1951), and others.

From Freudian psychology use is made of the concepts of repression and other intrapsychic defenses of "latent" and "manifest" content and of transferential and countertransferential manifestations, all as mechanisms of resistance to learning. The work of Caplan (1959) and Bernard (1964) on Public Health Psychiatry is useful, as are the recommendations of the Group for the Advancement of Psychiatry (1947-1956). Epidemiological, clinical, experimental, social and developmental psychology sources must also be brought into the context of an adequate training program.

REFERENCES

1. Freud, S.: Turnings in the Ways of Psychoanalytic Therapy; Collected Papers, vol. 2. Hogarth Press, London, 1924, pp. 401-402.
2. Surgeon-General's Ad Hoc Committee on Mental Health Activities and the Development of Comprehensive Health Programs in the Community. U.S. Public Health Service Publication No. 995, 1963.
3. Joint Commission on Mental Illness and Health, Action for Mental Health. Basic Books, New York, 1961.
4. Haylett, C. H., and Rapoport, Lydia: Mental health consultation. *In* Handbook of Community Psychiatry, L. Bellak, ed. New York and London, Grune & Stratton, 1964, pp. 319-339.
5. Lippitt, R.: Dimensions of the consultants job, in consulting with groups and organizations. J. Soc. Issues 25:5-12, 1959.
6. Mann, Edna: The Monitor Problem, Unpublished Community Mental Health Consultation Project. Postgraduate Center for Mental Health, February, 1953.
7. Moreno, J. L.: Who Shall Survive? Nervous and Mental Disease Monograph Series, No. 58. Washington, D.C., Nervous and Mental Disease Publication Co., 1934.

   ——: Psychodrama, vol. 1. New York, Beacon House, 1946.

   ——: The First Book in Group Psychotherapy. New York, Beacon House, 1932, pp. 7, 14, 15, 16, 72, 73.

   ——, ed.: Das Stegreiftheater. Berlin, Gustav Keipenheuer, 1923.
8. Reynolds, Bertha C.: Learning and Teaching in the Practice of Social Work. New York, Farrar and Rinehart, 1942.
9. Taft, J.: Function as the Basis of Development in Social Work Processes; Am. A. Psychiatric Soc. Workers, Newsletter No. 9, 1940.
10. Towle, C.: The Consultation Process. University of Chicago. Unpublished manuscript, 1950.
11. Robinson, Virginia: A Changing Psychology in Social Casework. Chapel Hill, University of No. Carolina Press, 1930.

    ——: The Dynamics of Supervision Under Functional Controls. Philadelphia, University of Penna. Press, 1949.

    ——: Training for Skill in Social Casework. Philadelphia, University of Penna. Press, 1942.
12. Austin, L.: Trends in differential treatment in social casework. *In* Principles and Techniques of Social Casework, Cora Kasius, ed. New York Family Service Association of America, 1950, pp. 324-338.

13. APTEKAR, H. H.: The Dynamics of Casework and Counseling. New York, Houghton Miflin Co., 1955.
14. MERTON, R. R.: Social Theory and Social Structure. Glencoe, The Free Press, 1947.
15. PARSONS, T.: The Structure of Social Action, 2nd ed. Glencoe, The Free Press, 1949 (1st ed., McGraw Hill, 1937).
16. HOMANS, G. C.: The Human Group. New York, Harcourt Brace & Co., 1950.
17. CARTWRIGHT, D., AND ZANDER, A.: Group Dynamics—Research and Theory. Evanston, Row, Peterson and Co., 1953.
18. BALES, R. F., STODTBECK, F. L.: Phases in group problem solving. J. Abnorm. Soc. Psychol. 46, 485-495, 1951.
19. Caplan, G.: Concepts of mental health and consultation—their application. In Public Health Social Work. Washington, Children's Bureau Publication 373, U.S. Department of Health, Education and Welfare, 1959.
——: Types of mental health consultation. Am. J. Orthopsychiat. 33:470-481, 1963.
20. Bernard, Viola W.: Education for community psychiatry in a university medical center. *In* Handbook of Community Psychiatry and Community Mental Health, L. Bellak, ed. New York and London, Grune & Stratton, 1964.
——: The Training of Mental Health Specialists in Schools on Public Health. Conference held at Arden House, N.Y. on Dec. 6-11, 1959, Sponsored by Association of Schools of Public Health, Columbia University, School of Public Health and Administrative Medicine.
21. Group for the Advancement of Psychiatry: Reports, Nos. 1-34, 1947-1956, New York.

Discussion by Albert Bryt, M.D.

I WISH to endorse the meritorious plea by Wolberg and Lawson that in our attempts at communication we should aim at defining our idiom so that it may be understood by the listener. My discussion, therefore, might be taken as a practical test of the possibilities of meaningful interchange between people coming from different disciplines.

Untold frustration may result for consultee and consultant alike when the latter's role is confined, no matter upon whose initiative, to that of making recommendations. Project B in the preceding chapter illustrates this unfortunate outcome. Most of us can be useful by providing the climate for a clarification of the consultee's thinking, putting at his disposal the specialized knowledge and insights which we have acquired in the course of our training and professional practice. Then, if necessary, with our assistance, such special insights may be adapted to the particular skills and background of the consulting personnel.

Many years of psychotherapeutic work with adolescent patients have made it necessary for me to enter into informal and not so casual consultative relationships with school authorities, law enforcement agents and institutions, as well as social agencies. When there was a great disparity between disciplines, a snag was likely to occur unless I was able to gear my formulations to the idiomatic level of my protagonist. In some instances I first had to provide a learning experience to the people who wanted my help before they could properly benefit from it. Project C reported by Mrs. Wolberg and Dr. Lawson would fit this latter experience, and should be scored as a success, albeit a limited one.

Consultant as well as "client system" must speak a language understandable to all. This may require modifications in psychiatric conceptualization; the authors and I share Ruesch's view that "psychiatric systems really make little sense: (that) they are primarily anthropological and historical museums of philosophical, psychological, and religious considerations of past centuries."[1] An egocentric orientation by the psychoanalyst creates a hazard in that it is likely to interfere with the necessary interpolation between experience with patients and that with consultees. We need to focus on intrapsychoanalytical modifications, so to speak, of an attitude which may be described in Grinker's words as "the self-complacency responsible for maintaining traditional attitudes and techniques. . . . Today, [he writes] we are asking searching questions about what kind of therapies are suitable to whom and whether these procedures are really effective."[2] "Psychoanalysis should be more greatly concerned with conditions of 'normality,' and with prevention of illness to warrant the trust placed in us by those from related fields who look upon us as their guides in treatment, and authorities in evaluating its success."[3]

The psychoanalyst must at least be conversant with the fundamental concept of cultural relativity which postulates that the same basic needs can be satisfied in a variety of ways, none necessarily better than the other.

Actually, such conceptualization is quite compatible with psychoanalytic thinking. As Parson has pointed out, patterned social interaction is not only a function of the motives of the individual but also "a function of the exigencies imposed by the coexistence of many relationships in situations where they are necessarily interdependent."[4]

Sullivan,[5] has laid the groundwork for divesting ourselves of our present diagnostic system, which holds us prisoners because it is designed to describe specific clinical entities in terms of likely psychopathology. In order to do justice to our task we must evolve terms which will measure levels of adaptation, delineate symptomatology and outcome, gauge the assets and the liabilities of the individual, the social group, and even the "client system." Our presently prevailing value-judgmental orientation leads us, in a given instance, to assess one particular area of adjustment in diagnostic terms. Community mental health must be viewed rather on a continuous adjustment-maladjustment scale with the focus on an entire lifespan within a cultural system, rather than on a particular chronological moment in one individual's existence.

Our common interests could meet in a frame of reference which would concern itself with life goals, overall adaptability and optimum functioning, taking into account psychodynamic and constitutional factors, social realities and cultural aspects, past handicaps and future possibilities. As consultants in community mental health matters we cannot afford to ignore the self-healing aspects of the personality nor the resources offered by others.

REFERENCES

1. Ruesch, J., and Bateson, G.: Communication The Social Matrix of Psychiatry. New York, W. W. Norton & Co., Inc. 1951, p. 66.
2. Grinker, R., Sr., MacGregor, Helen, Selan, Kate, Klein, Annette, Korman, Janet: Psychiatric Casework: A Transactional Casebook, New York, Basic Books, Inc. 1961, p.V.
3. Pollak, O., et al.: Social Science and Psychotherapy for Children. New York, Russel Sage Foundation, 1952, p. 29.
4. Parsons, T.: The Contribution of Psychoanalysis to Social Science. *In* Psychoanalysis and Social Process, J. Masserman, ed. New York and London, Grune and Stratton, 1961, p. 28.
5. Sullivan, H. S.: Conceptions of Modern Psychiatry. Reprinted from Psychiatry J. Biol. & Pathol. of Interpersonal Relations.

Part III. Training of Psychoanalysts in Community Psychiatry

THE RESPONSIBILITY OF PSYCHOANALYSTS IN COMMUNITY MENTAL HEALTH PROGRAMS

By PAUL H. HOCH, M.D.*

THE CONTRIBUTIONS of psychoanalysis to community mental health programs can be described under many headings. We can discuss the role of psychoanalysis in a hospital, a clinic, and in a private practice setting. We may also discuss the contributions of psychoanalysis to research and training in the framework of community psychiatry. It is very important that we go into the extent to which our present psychiatric treatments are adequate and where we need improvements or modifications.

There is a great deal of planning going on at present for community mental health services. It is my own concern that a great deal of time is devoted to the discussion of construction, organization, administration, coordination, and the relationship of psychiatry to other professions and other agencies. This, of course, is all necessary, but somehow evades the core problem. This problem is the application and effectiveness of our treatment methods. To assume that our treatment procedures are ideal and only the vehicles in which they take place are faulty is a gross misrepresentation of facts.

Those psychoanalysts who are fortunate enough to treat patients in the higher socioeconomic groups by a rather prolonged type of treat-

*Dr. Paul Hoch's death on December 15, 1964 will be long and poignantly regretted by his multitude of friends in the Academy and throughout the world.

ment leisurely and unhurriedly may have to stop and realize that there exists a large group of unfortunates who still need help. We do not want to maintain the cleavage that one of the highest trained groups in psychiatry treats only a few patients individually, but at the same time tries to give good advice to others how treatments should be provided.

Many of the conclusions reached in social psychiatry are still conjectural and many of the hypotheses which have been put forward to explain individual pathology based on social influences are unclear. Psychoanalysis will have the task of not only taking into consideration all our knowledge on the causation of emotional disorders, but also of dealing with the mental and emotional disorders in our society. A mental disorder may have an organic origin, and dealing with it can have many social implications. Psychoanalysis, regardless of its close relationship to the behavioral and social sciences is still a part of medicine. This has been stressed by Rado and Alexander. To quote the latter:

> "The physiologically determined development of the functions of intelligence, the maturation of the sex glands, senescence—these are all biological phenomena and constitute the most fundamental crises of emotional development. The individual has not only to adjust himself to a complex external environment, but also to an ever-changing internal environment which follows the unchangeable course of biological growth and deterioration. To try to understand the psychology of puberty or senescence without paying attention to the biological changes in the organism would be a procedure just as one-sided as to try to understand an individual with its complex structure of innate drives which operate in him without regard to external influences.
>
> We speak of executive functions of the ego and find its anatomical and physiological implementation in the motor apparatus. Is it not high time that we try more systematically to reconcile psychological with physiological data and abstractions and give up the traditional attitudes? While it would be nice if physiology were in accordance with psychological notions, yet we do not need to be seriously concerned with physiology in constructing a psychological model of the ego."

Redlich's definition of normality that a mentally normal person is an individual who does not need treatment by a psychiatrist is a practical one, even though we cannot consider it to be objective for the reason that clinicians often cannot agree on the criteria of mental health except in cases of severe illness. In its appearance physical disease is more or less absolute, whereas emotional disturbances can be constellative—present at one time and not manifest at another.

A further complication is the subjectivity of much of the emotional disturbance.

The community-based psychiatric services in other countries are trying to keep the mentally ill and, of course, the neurotics, in the community. They do not attempt this by curing them, but by relieving symptoms and adapting the patient to his environment and the environment to the patient. It can be said that in the United States psychoanalysts are interested in treatment and cure, whereas in Europe the emphasis is on social functioning and adaptation without much individual treatment. Both approaches have merit and a constructive compromise should be worked out. We feel that more clinical research should be devoted to distinguish between those who need short-term therapy and those who need prolonged psychotherapy, and to what extent other forms of treatment—somatic and social—could be effectively combined with psychoanalysis. It must be questioned if the patient has to be seen four or five times a week for a prolonged period of time or if it is possible to evolve treatments, which for a larger number of patients, could be made as effective by applying fewer sessions and for shorter periods. To quote Wolberg:

> "The shortening of therapy depends to a considerable degree upon the therapist himself. Unfortunately, there is a tendency to think about brief psychotherapy in terms of utilizing one or another special technique or frill. Actually, the most important factor in shortening psychotherapy is not the method that is used, not the specific syndrome treated, but the therapist himself—his understanding of dynamics and how to use himself most constructively in the therapeutic interpersonal relationship. What this means is that an incompetent therapist's techniques are not purely enhanced through the utilization of short-term approaches." I may also add that short therapy is not therapy administered by the unskilled, the amateur, and the unsophisticated; again it will depend who is treated and for what kind of condition.

I may also quote Masserman who correctly states that some orthodox analysts, if they are forced to do "brief psychotherapy" by the exigencies of practice, still fail to utilize the variegated resources of a truely dynamic psychiatric armamentarium and persist in attempting "abridged analysis," the sole objectives of which are to impart verbal "insight" into the "psychogenesis" of the illness as rapidly as possible. Here too, the tendency is to merely "interpret" the patient's presenting difficulties in terms of his remote or infantile past instead of regarding aberrant development simply as a background for his current problems and maladaptations which are amenable to active and constructive amelioration by every means at the command of the patient's physician, family, and community.

We will have to examine the extent to which psychoanalytic theories and psychoanalytic treatment apply to community psychiatry. Psychoanalytic theories about human development, sexuality, the unconscious, and the ego forces have validity just as much in the community setting as elsewhere. Some of these concepts could be applied to the interaction between persons and groups and some dynamic generalizations could contribute to the organization of community services, provided that psychoanalytic therapies do not remain isolated but are combined with biological and social factors. Furthermore, it does not mean that the validity of some of the psychoanalytic theories about human behavior have to lead automatically to the application of psychoanalytic therapy in every case. Psychoanalytic therapy and practice are, of course, inter-related, but we need not insist on an exclusive technique that in the last years has undergone considerable questioning and has led to new formulations. (Horney, Kelman, Sullivan, Rado, Alexander, Masserman, and others).

Not enough attention has been paid to the role of the superego, and more research must be done regarding the constructive role of society in formulating ethical, moral, and other concepts, and in turn, how they could and should become a part of the ego regulative functions. One of the difficulties in our society today is that we are confronted with an utter confusion of superego values which then lead to very conflicting treatment approaches and social visualizations of psychiatric and psychoanalytic attainments. To return to Alexander:

> "The thesis that psychoanalysis is a method which helps the individual accept his environment is more than an oversimplification. It is erroneous because it neglects one fundamental aspect of psychoanalytic treatment. Psychoanalysis aims at something more than merely helping the patient adjust himself to the social system in which he lives. Psychoanalysis is a true product of that phase of Western civilization which had deep respect for individual differences that people actually have in our complex free society. It aids a person in the integrative task in which he failed to reconcile his own basic personality with his environment without sacrificing that intangible something which makes him a person different from all others. Psychoanalysis is the most individual type of treatment medical science has ever produced. Each case is a unique problem. Psychoanalysis aims at an adjustment which takes cognizance not only of a given social environment, but also of the uniqueness of every person. The respect for individual differences explains why psychoanalysis is unthinkable and therefore prohibited in a totalitarian society."

Psychoanalysis in community psychiatry can make the great contribution of counterbalancing the trends to mass-apply different psychiatric techniques. The individual will still retain his unique strivings, shortcomings, struggles, and adaptation, and his treatment should not be submerged completely in social manipulative attempts, regardless of what methods of programming we follow in general rehabilitation.

Research and training must be devoted to the special requirements of psychiatric practice in a community setting, i.e., the application of psychoanalysis in a new and imaginative way in a community. One of the priorities should be to define more accurately those who need reconstructive and those who need supportive treatment. In some places the efficacy of supportive treatment is undervalued and therefore patients are rejected for treatment because it is assumed they cannot be reconstructed. On the other hand, in many places, every individual regardless of his symptomatology, his personality organization, or social setting is assumed to be a candidate for psychoanalytic treatment. The main difficulty here is that at present we do not have enough reliable instruments in our hands to determine which patient would benefit from a more intensive form of treatment and which patient could go along with much less. The indications for individual or group therapy are also based largely on the personal predilection of the therapist or the setting in which he operates—private office, clinic or hospital, etc.—but not on what would be the best for the particular patient. Psychoanalysis also will have to investigate more closely the relation of aggression to antisocial behavior, and not eliminate from treatment delinquents because it is assumed *a priori* that they will not respond to it.

The evaluation of psychotherapy and psychoanalysis has not been attempted sufficiently, frequently because it has been assumed that so many factors play a role in a change of the individual's emotional life that we cannot evaluate all of them adequately. It is true that any evaluation becomes more difficult the more factors there are to be considered in relation to cause and effect. Nevertheless, evaluation of psychoanalysis is possible if the process were to become divided into component parts and each of them studied from a methodological point of view. Improvements are judged by different psychiatrists in different ways because there is no agreement as to criteria. One judges a patient's improvement on social adjustment, another on sexual adjustment, and a third on insight; one uses descriptive terms, another psychodynamic formulations, which again vary with the different schools of psychodynamics. Even if we were to have considerable difficulties in evaluating component parts of psychoanalytic therapy— transference, resistance, etc.—this still should not prevent us from establishing a valid methodology with which outcome could be measured. We do not have objective instruments of measurements for many of the subjective manifestations in psychoanalysis, and some phenomena will always remain subjective and resist objectivation. Nevertheless, no therapy can make claims of accomplishments when it does not offer the possibility of scientific appraisal of its outcome.

It is important that training psychiatrists, general practitioners, and other medical specialists in psychoanalytic principles should be done clearly in understandable language free from contradictions and ambiguities. It should also clearly indicate what is known and what remains unknown or is in a state of investigation or review. A differentiation should be made in the training of this group between psychoanalytic techniques applicable only to a few and those psychodynamic formulations which could be used in any form of psychotherapy or combinations with somatic therapies.

PSYCHOANALYTIC TRAINING AND COMMUNITY PSYCHIATRY

By ROBERT G. HEATH, M.D.

THE PUBLICITY given community psychiatry has caused the public to consider it somewhat wishfully; as something distinctly new and different to eliminate mental illness. Community psychiatry actually is a special approach evolving from the need to treat large numbers of patients with limited manpower. A number of medical school departments of psychiatry are attempting to teach techniques of community psychiatry which, like all psychotherapeutic techniques, are based on our knowledge of human behavior.

Community mental health programs require broad application of knowledge gathered from intensive study of small numbers of patients with behavioral disorders. As a consequence of recent progress in our specialty, methods have presumably evolved for more rapid, effective management of large numbers of mentally ill persons. Experiences in a few centers indicate that energetic, imaginative community programs can indeed be formulated from current knowledge.

Emphasis on community mental health is the latest of several major shifts of direction in academic programs for training psychiatrists. Immediately before World War II, diagnosis and classification of mental illness received primary attention, and psychotherapy was largely supportive. Then psychoanalysis began its ascendancy, stimulated by frustration and by optimism that this technique would provide the foundation for more complete understanding and, therefore, prevention and treatment of mental illness. Many of us believed that psychoanalysis was the ideal therapy and that, started early enough, it would prevent almost all mental disorders. As a result, many university residency training programs were shaped on the psychoanalytic model, emphasizing an intensive, prolonged one-to-one doctor-patient relationship. Emotionally charged side issues developed as some academic institutions attempted to integrate psychoanalytic training with residency training. From World War II until a few years ago, the "dynamic," or psychoanalytic type of residency training programs, were the popular choice of physicians who were entering the specialty. But with the gradual emergence over the past few years of a more eclectic approach and with the realization of our community responsibility as physician-specialists, a new identity for the psychiatrist has begun to emerge.

The framework for the study of human behavior is usually psychodynamics, which, by definition, represents a theory based on data gathered primarily by psychoanalysis but by other techniques as well, whenever possible. Dr. Rado, in 1946, advocated the teaching of psychodynamics as a basic science,[5] and recently developed this concept in describing the relation of adaptational psychodynamics to therapeutic programs in community psychiatry.[6]

Shortly after the Department was organized at Tulane in 1949, we completely integrated training in psychoanalysis into the psychiatric residency program.[2,3] Our goal was to train the resident in effective use of a large number of diagnostic and therapeutic procedures; to develop his competence in physiological and pharmacological treatments, and to use a broad range of psychotherapeutic techniques in contrast to a single technique, as is customary in psychoanalytic institutes. Our basic model for training, however, has remained the classical one-to-one doctor-patient relation derived from the psychoanalytic approach.*

The major training exercise for our residents is outpatient clinical experience in intensive psychotherapy.[4] On this service, the resident sees a small number of patients frequently and over a relatively long period. Instruction is given through individual supervision, supplemented by lectures and seminars. This model was first established in the psychoanalytic institutes. Even in view of the new emphasis on development of techniques for treating large numbers of patients, we consider our method of teaching the psychiatric resident to be fundamental and sound, as long as he does not consider it the endpoint in therapeutic technique. We have increasingly emphasized to the resident that the experience in intensive psychotherapy is primarily a teaching exercise which provides the best opportunity for understanding psychopathology and psychodynamics. We consider the resident's personal psychoanalysis in this same light. Although personal psychoanalysis is optional in our program, it is elected by most trainees. The knowledge

*Tulane University gives certification in psychoanalytic medicine to the resident who meets the requirements of the program.

of psychodynamics gained from training in intense interpersonal relations, through personal analysis and intensive psychotherapy, provides an excellent basis for the development of other therapeutic methods dictated by clinical demands.

The influence and position of psychoanalytic training in our program have changed since 1949. In the early years of the Department, most of our residents were primarily interested in becoming certified practicing psychoanalysts, whereas today few are interested in pursuing psychoanalysis exclusively. Since 1957, in fact, none of our residents has expressed interest in psychoanalytic training in an institute affiliated with the American Psychoanalytic Association, although New Orleans has such an institute with some of our faculty as members. Instead, most of them apply the knowledge gained through psychoanalytic training in a widely diversified way. Neurophysiological experiments on the brain, for example, have been conceived and interpreted on the basis of psychodynamic theory. Similarly, new psychotherapeutic techniques have evolved from this basic theoretical information. Exposure to experiences in community psychiatry through new teaching exercises have interested more and more graduates to use psychodynamic principles in preventive techniques.

Further evidence of a shifting interest is the fact that ten years ago almost all our residents entered full-time or part-time private practice after completion of training, whereas none of our twelve graduates in 1963 entered practice; instead, all assumed full-time academic or community service positions. Of the thirteen residents who completed psychiatric training in 1964, the only one who entered private practice is also part-time director of a community clinic and teaches one day a week in our Department.

Development of Community Psychiatry Training Programs

To those of us responsible for training psychiatrists, the important question is: How can we make sure that residents who complete psychiatric training are competent in using specialized technics in community psychiatry? More specifically, how can the resident, trained in the one-to-one therapist-patient model of the clinic, apply this knowledge to help large numbers of sick or potentially sick people in a community setting? The final answer to this question awaits further experimentation. Although programs in community psychiatry are now operating in several schools, experience is still too limited to provide definitive answers.

A variety of public health approaches to psychiatry are being used in these centers; some have added psychiatrists to the faculties of schools of public health (Johns Hopkins, Columbia), basing the program on public health, whereas others have appointed public health specialists to the faculties of departments of psychiatry in the medical schools (Harvard and the University of California at Los Angeles). At Tulane, we have developed one program based in both schools, the School of Public Health and the Department of Psychiatry of the School of Medicine; the teaching faculty holds dual appointments. Although the program has been operating for several years, we still consider it to be exploratory. Our basic aim is the integration of the community-oriented approach with our dynamic psychiatric program, specifically, the application of psychodynamic knowledge by means of public health methods. The two foremost questions confronting us in the development of this program are: (1) What body of information should be taught? and (2) How should it be taught? In organizing our program, Dr. James Knight stipulated that the instructional material include, in addition to conventional clinical training, at least the following nine nonclinical aspects of practice.

Content of Instructional Material

1. *Orientation in community organization.* A knowledge of community organization is essential to the most effective application of psychodynamic theory to the prevention and alleviation of behavioral disorders. The trainee should learn the functions of the various community agencies, how they are interrelated, how information is transmitted from administrators to workers to the Board and finally to the community, and how the needs of the community, in turn, are transmitted to the agencies.

2. *Program administration.* The trainee should learn methods of planning, managing, staffing, and financing all kinds of mental health services. Neglect of this aspect of training in the past has resulted in a shortage of able administrators for programs of community psychiatry.

3. *Mental health education.* The trainee should gain competence in use of mass media and other special educational techniques he will

later teach, as a community mental health specialist, to health educators, lay leaders, and specialists.

4. *Consultation.* The techniques of consultation are usually not emphasized in psychiatric residency training programs. Yet, as Dr. Gerald Caplan particularly has demonstrated, one of the most effective methods of applying psychodynamic principles to treatment is psychiatric consultation to nonpsychiatric agencies and professions.

5. *Applied research.* With the inception of community psychiatry programs, the most meaningful research will logically be conducted first in the field of epidemiology, one of the four principal courses taught in the schools of public health. This science is fertile for experimentation, since it is just now being applied to psychiatry. Biostatistics, another major course in schools of public health, is invaluable to the community mental health specialist in evaluating the effectiveness of new psychiatric programs.

6. *Forensic psychiatry.* Greater emphasis should be placed on forensic psychiatry in the residency program. The specialist in community psychiatry, in assuming greater responsibility for large numbers of citizens, must be aware of laws that preserve the maximal rights and liberties of the mentally disabled consistent with the physical welfare of the patient and the community.[7]

7. *Group dynamics, group psychology, dynamics of leadership, communication theory and practice* are essential fields of knowledge in the implementation of community psychiatry programs, and should receive special attention during training. An awareness of interpersonal and intergroup dynamics is the stock-in-trade of the clinician, but he uses this knowledge chiefly in a limited therapeutic context. An understanding of the dynamics of human relations, however, can also be used advantageously in daily negotiations with community groups.

8. *Supervision.* Individual supervision of the resident's training in psychotherapy is the cornerstone of most dynamic residency programs. For training in community psychiatry, similar supervision should be provided for the resident's activities with community groups.

9. *Special methods for inservice training.* The trainee should be taught how to organize educational programs and workshops in mental health for nonpsychiatric professions, for staffs of psychiatric as well as nonpsychiatric agencies, such as welfare, public health, or probation departments, and for schools.

Methods of Instruction

Instructional methods for residents vary in different training settings. Traditionally, training in child psychiatry has been the most community-oriented aspect of the residency program, since it was believed that elimination of environmental stresses would correct the behavioral disorder before a fixed neurotic pattern was established. Increased interest in community mental health has led to expansion of this aspect of child psychiatry training.

Programs emanating from schools of public health have also contributed to the development of community psychiatry. In some centers psychodynamic knowledge has been fused with that in the major subspecialities of public health, epidemiology, biostatistics, public health administration, and environmental medicine and hygiene.

Community Psychiatry Training at Tulane

At Tulane, training in community psychiatry is being given in several parts of our program: during the three-year residency in general psychiatry, during career training in child psychiatry, and in a special program of training for specialists in community psychiatry.

Second Year of Residency

Exercises in community mental health begin in the second year of residency. Our intent is to orient the trainee in community organization and provide him preceptorship experience in consultation. The second year trainee participates as a preceptor in activities of civic and governmental agencies, primarily in the city of New Orleans. For example, he attends meetings of the City Council, of the Poverty Program, and of the Professional Advisory Committee, Staff and Board of the Social Welfare Planning Council. He visits the Parish Prison and the Courts, attends police line-ups and police staffings, and rides in a patrol car of the First District of New Orleans. Once a week throughout the year, the second year trainee attends a conference conducted by Dr. Edward Norman, director of our community psychiatry program, at which time these experiences are reviewed.

Third Year of Residency

During the third year of psychiatric residency, each trainee spends one day a week at a State-operated community clinic, where he conducts diagnostic and therapeutic interviews with patients referred to the clinic from various sources. His work is supervised by the psychiatric staff of the clinic and visiting members of the full-time faculty of the Department. In addition to treating a limited number of patients, the trainee supervises members of the nonpsychiatric staff of the clinics and consults with representatives of community agencies. On occasion, the resident is asked to speak before a local community organization, such as the Parent-Teachers Association or a church group. He also becomes acquainted with the general administrative policies and problems of the clinic. In appropriate circumstances, he is invited to join the director in handling administrative problems. He becomes acquainted with the clinic Board members and, whenever possible, attends Board meetings.

When first assigned to a clinic, the resident is introduced to the important agencies and institutions in the community, including the local governing body, law enforcement agencies and courts, local State-welfare agency, local public health department, schools, churches, and private social agencies. As he acquaints himself with the community, he enters into more extensive consultative activities and undertakes some primary preventive measures. He is encouraged to evaluate any changes that occur within the clinic or the community during his twelve months' assignment.

All third year trainees participate in the weekly conference conducted by Dr. Norman in the Department. The residents summarize the activities of their clinics, including a description of the community, relations with clinic staff, supervision, and consultation with community agencies.

During the half-time that the third year resident spends on our child psychiatry service, he gains preceptorship experience in consultation at various facilities; the rotational schedule is such that he visits one facility each week for approximately eight weeks.

1. *Crippled Children's Hospital.* This is a client-centered[1] consultation, which includes diagnosis of the handicapping condition, medical, surgical, and corrective care, facilities for hospitalization, and after-care.

2. *Magnolia School, Inc.* A consultee-centered and client-centered consultation is conducted at Magnolia School, Inc., a nondenominational, nonprofit day-and-boarding school which provides training for handicapped children, particularly those with emotional or personality problems, speech difficulties, convulsive disorders or physical incoordination, and "improvable" mental retardation.

3. *Milne Boys' Home.* Client-centered and consultee-centered case consultations are conducted at Milne Boys' Home, which provides boarding care for boys between eight and sixteen years of age who need training and rehabilitation.

4. *Youth Study Center.* Client-centered and consultee-centered consultations are held with the Youth Study Center, a detention home for boys and girls between eight and sixteen years of age. Social services for the Center are provided by the probation staff and the juvenile court.

Career Training in Child Psychiatry

Residents undertaking career training in child psychiatry have more intensive and prolonged training in consultation. During the latter part of their training, in addition to attending the facilities already listed, they function as consultants, under the supervision of faculty, to the following community organizations:

1. *Associated Catholic Charities.* The Associated Catholic Charities provides casework service for nine child-care institutions, for foster homes, unmarried mothers, and adoption agencies. These consultations are scheduled once a month, but additional ones are provided on request. The consultation includes client-centered and consultee-centered cases and discussion of administrative concepts.

2. *Visiting Teachers* (*Orleans Parish School Board*). The Visiting Teachers provide casework services to children who show maladjustments in school and children with attendance problems. Consultations, which are patient-centered and consultee-centered, are scheduled weekly by the career trainees in child psychiatry.

3. *Juvenile Court* (*Orleans Parish*). The career trainee, under supervision, serves as a consultant to the Juvenile Court once a month. The Court has jurisdiction of proceedings concerning delinquent or neglected children under seventeen years of age. Consultations are client-centered and consultee-centered.

4. *Jewish Family and Children's Agency.* Once a year consultee-centered consultation is provided this Agency, which affords casework services to families and children. Additional consultations are available on request.

5. *Children's Bureau.* Consultee-centered consultations are furnished on request to this Bureau, which provides casework counseling for parents when the children's problem is the chief concern.

6. *Family Service Society.* On request, the career trainee, under

supervision, provides consultee-centered case consultations to this Society, which offers casework counseling to families experiencing difficulties in family relations.

Career Training for Specialists in Community Psychiatry

A career training program for specialists in community psychiatry has been initiated by the Department as one of its extension training programs. Eligibility for the program includes completion of the basic three years of psychiatric residency. Since our trainees at this level are often working toward certification in psychoanalytic medicine, they may simultaneously be training in psychoanalytic medicine and community psychiatry.

This program for specialists operates in collaboration with the School of Public Health of the University and leads to the Master's degree in Public Health. The work may be taken over one or two years. The curriculum is primarily public health oriented, with emphasis on epidemiology, biostatistics, public health administration, and environmental medicine and hygiene. Additionally, training exercises are designed to assist the resident in integrating the methods used in public health with his training in dynamic psychiatry. As part of his work toward the Master's degree, the trainee is required to complete a thesis on some aspect of community mental health that meets the approval of both the Department of Psychiatry and Neurology and the School of Public Health.

Summary

In the residency training program in community psychiatry at Tulane, psychodynamics is considered a basic science of human behavior, from which numerous therapeutic and research techniques evolve. Residents are trained in methods of applying certain specific techniques to the treatment of large groups of patients. Experience in nonclinical aspects of community psychiatry, including administrative problems is an integral part of the program.

REFERENCES

1. Caplan G.: Types of mental health consultation. *In* Principles of Preventive Psychiatry. New York, Basic Books, Inc., Publishers, 1964, pp. 212-231.
2. Heath, R. G., Monroe, R. R., and Lief, H. I.: The integration of psychiatric

and psychoanalytic training at Tulane: A ten year overview, 1950-1960. *J. Med. Educ.* 36:857, 1961.

3. ——, ——, and ——.: The integration of psychiatric and psychoanalytic training at Tulane: The psychoanalyst as teacher and administrator. *In* Science and Psychoanalysis, vol. V., J. Masserman, ed. New York and London, Grune & Stratton, 1962, pp. 216-227.
4. Lief, H. I., Lief, V. F., Warren, C. O., and Heath, R. G.: Low dropout rate in a psychiatric clinic. Arch. Gen. Psychiat. 5:108, 1961.
5. Rado, S.: Psychodynamics as a basic science. Am. J. Orthopsychiat. 16.405, 1946.
6. ——: Scientific psychotherapy for the millions. Presented at the 120th meeting of the American Psychiatric Association, Los Angeles, May, 1964. In press.
7. Lindman, F. T., and McIntyre, D. M., Jr., eds.: The Mentally Disabled and the Law. Chicago, University of Chicago Press, 1961.

TRAINING FOR COMMUNITY MENTAL HEALTH SERVICES

By JOHN A. P. MILLET, M.D.

WHILE ENGAGED in writing this chapter my attention happened to light upon an article in the New York Times of Sunday, August 23rd, 1964, which reported a vitriolic attack on American psychiatry as being completely dominated by psychoanalysis and isolated in its conceptual framework by this fact from the general stream of psychiatric knowledge and practice as observed in other countries where the specialty is "more advanced." The writer, William Sargant, a British specialist who favors the mechanistic and pharmacological approaches to therapy, had part of his graduate experience in this country, and was an honored guest of the American Psychiatric Association in Los Angeles last May. His attack, according to the Boston Globe, evoked a concerted angry rebuttal by leaders of psychiatric education such as Jack Ewalt, Peter Knapp and Bernard Bandler. While this attack was obviously laden with prejudice it should awaken us to facing some very important questions. Are we as psychoanalysts in danger of becoming too absorbed in our own special ways of thinking about people, of becoming isolated from the stream of progress in the struggle to better human relationships and combat the sources of mental illness? Are we then abdicating all claim to be workers in the community vineyard? If not, wherein lie our competence and claim for recognition?

I shall attempt to answer these questions. What I will have to say is in part a further elaboration of a thesis which I hold in common with many others, I suspect, that it is time for a change in the image which the psychiatrist, and especially the psychoanalyst, holds of himself, and which he finds now seriously distorted in the vision of many groups in the community. I expressed some views on this subject in a paper which I contributed to the panel on Social Aspects of Psychiatry at the recent international congress in Mexico City.[1]

While there is no need to waste time and space in discussing the semantics of terms commonly employed in the literature there is one specialty, social psychiatry, which is by some considered to include the subspecialty of community psychiatry. An excellent summary of the reasons given for this breakdown of psychiatrists engaged in community work into two categories is provided by Dr. Leopold Bellak's Handbook of Psychiatry.[2] My own view is that the social psychiatrist is the sort of community psychiatrist who has special interest in using part of his time and energy in studying the composition of special groups of human beings in collaboration with social scientists, psychologists, and others, making use of the insights of group dynamics. In this view, I think I am in essential agreement with Dr. Viola Bernard.[3] Such work is of the greatest importance as background for establishing the sort of educational, preventive, and therapeutic programs that are likely to be successful in the given community.

The advance of the community mental health movement was abruptly halted with the outbreak of the second world war, only to be renewed with deeper conviction and enthusiasm by its leaders when the task of rebuilding disrupted societies became the first task of government and citizen alike. The depression that shrouded the sessions of the Second International Congress on Mental Health in Europe in 1939 was effectively dispelled in 1948 when the World Federation for Mental Health was founded in London.

Fortunately, during the war, many of the younger psychoanalysts had risen to positions of great responsibility in directing programs of treatment for psychiatric casualties at home and abroad. At one time all three branches of the armed forces had psychoanalysts at the head of their psychiatric programs. Their experiences and the influence which they brought to bear on many young physicians under their command was responsible not only for the great multitude of applicants for psychoanalytic training after discharge, but also for the continued interest of those psychoanalysts who had held command posts in maintaining contact with community work when they resumed their civilian responsibilities. Due largely to their efforts, leadership in the American Psychiatric Association was transferred from the group of hospital directors to psychodynamically oriented psychiatrists, many of whom were psychoanalysts.

For a time, this tendency toward a broadened area of interests was looked upon with disfavor by some of the more constricted and rigidly trained psychoanalysts. Such interests were thought likely to interfere with the detachment considered necessary for scientifically precise work in psychoanalytic therapy. The pure gold of psychoanalytic dedication might become debased by the dross of involvement in action programs. Today, however, the American Psychoanalytic Association has been aroused to the danger of losing prestige if a large proportion of its members cannot be shown to be participating in teaching, research, or some form of community work. Not the least important of these special functions is that carried by instructors in medical schools, for it is their influence as participants in the clinical teaching of the undergraduate medical student that does most to take the mystery and nonsense out of what is too often said about psychiatry and psychoanalysis in academic quarters.

It is clear then that psychoanalysts in our country have for the past generation spearheaded the campaign for promoting significant changes in the education of physicians, in the care of hospitalized patients, and in devising programs designed for the prevention of breakdowns in a wide variety of special facilities. In fact, psychoanalysis as the basic discipline, and the emerging science of psychodynamics, which is rooted in its concepts, have been thought by many leaders of the profession, notably those charged with setting the standards for certification by the American Board of Psychiatry and Neurology, to have favored too one-sided a view of what qualifies a man to be certified for competence in the specialty of psychiatry. There is a growing tendency to doubt whether recent graduates from a psychoanalytic Institute have mastered or are even basically interested in the fundamentals of clinical and community psychiatry. A growing percentage of applicants for certification who have recently completed their psychoanalytic training have been turned down, including not a few who hold positions on psychiatric faculties as lecturers and clinical teachers.

It is hard none the less to see how any group could be better fitted, theoretically at least, to take effective roles in developing and directing programs of prevention and care in the fields of mental health and mental illness than fully trained psychoanalysts—always provided that they are not the sort whose interests and skills lie in some other direction such as psychophysiological research. Understanding of the fundamental principles of psychodynamics is a prerequisite, it would seem, for effective participation in all such programs, though in this field as in any other where the success of an undertaking is dependent on the quality of the interpersonal relationships involved in the enterprise, what counts most is the sensitiveness to other people's feelings and to the roles of his nonmedical colleagues on the part of the leader rather than the number of certificates he can hang in his office.

Psychodynamics as the science of motivation, understanding and control can be learned by rote, but intellectual grasp of the principles involved is no guarantee of their consistent or even adequate implementation. We are now faced with an impressive challenge. For the first time the nation is quite fully aroused to the threat of mental illness to the economic and social health of the national community. New overall programs devised to attack this danger at every level where its presence is suspected or recognized are in the planning stage. Much of the impetus toward planning for community mental health stemmed from the work of the GAP Committee on Social Issues. Their thinking was an important factor in helping the Joint Commission on Mental Illness and Health to formulate their final report (Action for Mental Health).

New York State was the first state to enact comprehensive legislation aimed at awakening communities to the problem and to offer a helping hand in financing any suitable plans which they could develop. The Community Mental Health Services Act of 1954 made it possible for every county or municipality to establish a community mental health service, with the proviso that if adequate quarters were secured and properly qualified staff appointed the state would match the appropriation dollar for dollar. I had the privilege of helping to establish the service in Rockland County, where we were the sixth county to gain state approval. Serving first as chairman of the professional advisory committee of the local mental health association, then as first executive director of community mental health services, and finally as chairman of the mental health board, I had the opportunity of experiencing many problems involved in getting such a service started in a normally resistant community of 100,000 people and in seeing the beginnings of a well-organized comprehensive community mental health service. The problems faced by anyone who attempts to pioneer in this field vary in all kinds of details, but almost always have some features common to all, notably fears on the part of the medical profession that this is just another step leading to socialized medicine, and the fears of other community agencies such as the school system, in some instances, or, in others, the courts, that their established prerogatives might be threatened with invasion from a new source. On the other hand, in some instances, it may be just one of these agencies—in the case of Rockland County, the school system, as represented by the school psychologists—which proved to be the chief promoter of the enterprise.

Since the Joint Commission on Mental Illness and Health came out with its report the American Medical Association began to pay more serious attention to its national responsibility for helping to implement some of the more important recommendations of the Joint Commission in the field of community mental health. In 1962, a call was issued to all agencies, governmental and private, interested in the training of mental health personnel and the planning of community mental health services, to send representatives to the First National Congress on Mental Illness and Health. At this congress there was a great deal of discussion about the training and use of nonprofessional lay personnel as auxiliary aides in such areas as group discussions and individual counseling. Many psychiatrists who attended the meeting came away with the feeling that there were real dangers involved in letting down standards of care which had been developed through years of experience by teams of professional workers. The paucity of experts trained in community work, however, in all three of the specialties principally identified with mental health work, (psychiatry, psychology and social work) when considered against plans for a massive attack on mental illness at the community level, seemed to most thoughtful people to require imaginative planning in a new and enlarged dimension.

Having this information at our disposal we must ask ourselves several leading questions:

First: What is our responsibility as physicians and psychoanalysts for assisting in the planning and implementation of community mental health services?

Second: What position if any should we take with respect to the definition of our special role and those of other professionals and untrained laymen in developing a philosophy and program of action for mental health?

Third: What opportunities are there for training of psychiatrists to take leadership roles in the field?

I would like to address myself first to the main question. What is our responsibility as physicians and psychoanalysts? I have already indicated my view that we need to develop a new image in order to assure ourselves an honorable part in the overall attack on mental illness. Psychoanalysts, as I pointed out, have been among the foremost leaders in the planning and execution of programs offering preventive and therapeutic services to their communities in peace and in war. As examples we might mention Jules Masserman in Chicago, Erich Lindemann in his Wellesley project, Portia Bell Hume in San Francisco, and in New York State Drs. Louis C. English and Margaret M. Lawrence in Rockland County, Dr. Leopold Bellak in Queens, and Dr. Lawrence Kolb in Manhattan. Others have taken leadership roles in the making of surveys of communities from the standpoint of social psychiatry, such as that conducted by Dr. Alexander Leighton in Stirling County and the Washington Heights survey under the direction of Dr. Lawrence Kolb.

From the choice of these examples you will see that all those mentioned have started as clinicians, soon to become teachers, and then have developed wider interest in the significance of their special knowledge to the study and solution of mental health problems at the community level. At the same time, each has pioneered from a different base of operations—Masserman from leadership as teacher and research worker, Lindemann in a pilot project in family relations, Hume as director of a state mental hygiene clinic, English and Lawrence as director and assistant director of a local county all-service mental health clinic, Leighton from his position as Professor of Social Psychiatry at Cornell, and Kolb as head of the Department of Psychiatry at Columbia. It is my earnest conviction that in the future all applicants for psychoanalytic training should either show proof of having had basic experience at the community level or should be required to spend at least one semester in either a general hospital with both inpatient and a variety of outpatient services, or in one of the regional centers for mental health envisaged in present national plans. Whereas most of us are now involved in some phase of clinical work at the community level, those others whose interests do not lie exclusively in the area of research would be well advised to devote part of their time to some community service.

Then as to the second question. It is a cardinal principle of all successful participation by psychiatrists in mental health work that they should approach their tasks with a proper attitude of scientific humility, aiming first at helping to study the situation and the roles which they are expected to fulfill, conferring with their colleagues in other disciplines and/or on other medical services in the spirit of joint consultation, are thus recognizing the magnitude of the task ahead of them and the fact that success in attacking the problem is dependent on the skill used in developing the sort of understanding collaboration that is indispensable for effective team work. The need for such an approach would seem obvious; the understanding of the processes of group interaction which has been provided by students of group dynamics should be made available to all students of psychoanalysis. Work at the community level is always dependent on good team work, and the psychoanalyst who has sensitivity to the subtleties of human interaction should be the ideal captain of such a team. Such a captain cannot be expected to be personally expert in every detail of the team operation, but should become so aware of the distribution of functions among his team members, and of the strengths and weaknesses of each, that his role becomes essentially that of an overall consultant and team coordinator for which his special expertise as a well-trained clinician with proved psychodynamic insights qualifies him.

Finally the third question. What are the opportunities for specialized training and education for community psychiatry? I shall here mention a few which are offered by institutions in which psychoanalysts have taken leading roles. Thanks to the influence of a succession of leaders in the organization of psychiatric education at the Harvard Medical School and in the Boston community, the Harvard School of Public Health established a Division of Mental Health where the leading professional positions have been held both by psychiatrists and, as at present, by Louisa P. Howe, a distinguished sociologist who is an affiliate member of the Boston Psychoanalytic Society. The Mental Health Consultation Project in which she participates is sponsored by the Harvard School of Public Health. This school has pioneered in offering graduate education for community psychiatry.

In 1957, California developed a program for community mental health services based on the New York State Community Mental Health Services Act of 1954. Since then the Department of Mental Hygiene has established a center for training in community psychiatry of which Dr. Portia Bell Hume has been director since 1961. This center offers eight courses to mental health workers (psychiatrists, social workers, nurses, psychologists) who are basically clinicians, with a minimum of two years field experience after receiving their Master's degree (social workers and nurses), or their Ph.D. (psychologists).

Psychiatrists are accepted after completion of two years approved residency. Each of the eight courses lasts seventeen weeks. There is also a summer session of four weeks, full time.

In 1956, a Division of Community Psychiatry was established at Columbia University under the joint auspices of the Departments of Psychiatry and the School of Public Health and Administrative Medicine. Dr. Viola Bernard was appointed director of the new curricula established under this combined leadership. Enrollment is limited almost exclusively to psychiatrists. The five curricula offered are designed to meet a variety of needs. The core curriculum includes courses, seminars, and field trips, supplemented by reading seminars, lectures, and discussions on every conceivable aspect of psychiatry in its application to the community. This curriculum calls for a four-year psychiatric residency and leads to Board eligibility in Psychiatry and the degree of Master of Science or Master of Public Health. For qualification in child psychiatry the curriculum calls for an additional year of residency.

In addition to these two basic curricula the Division offers a two-year postresidency training in community psychiatry leading to the degree of Master of Science or Master of Public Health, and two shorter curricula; one, administrative psychiatry, calling for one year, full time, leading to the degree of Master of Science, the other a fourth year residency training program, or part-time postresidency, in community psychiatry, leading to no degree.

We can see here then two types of programs, one originated by a Department of Mental Hygiene (California), the other by joint faculties in an University School of Medicine (Columbia). The difference in emphasis is specially interesting, in that in the former full recognition is given to the need for quite thorough training of professional groups other than psychiatrists, thus emphasizing the collaborative nature of work in the field of community mental health, whereas in the latter the emphasis in the program is on creating curricula designed for the benefit of psychiatrists, with recognition of the great need for preparing psychiatrists to become leaders in community psychiatry and/or social psychiatry. Due recognition is here given also to the necessity of educating administrators with a well-rounded psychiatric background in addition to those more inclined toward the development of preventive and therapeutic services or in doing research. It is worthy of comment that in both instances the programs were developed and directed under the leadership of psychoanalysts. The foresight and thoroughness which characterize the organization of these and other new educational programs provide a ready answer to the criticisms of psychoanalytic influence on American psychiatry offered by our British colleague.

If we take a hard look at the broad canvas of functions which are subsumed under programs for development of comprehensive mental health services in the community it would not be hard to find some special area or areas of professional interest and competence having appeal to each of us. The community psychiatrist does not have to be a skilled and experienced expert in every phase of functioning included under the heading of community psychiatry. He may have a special interest in different forms of therapy, for example, and be a specially skillful clinical supervisor. He may have a preference for giving the administrative function first importance. He may have been a special scientist or a psychologist before becoming a psychiatrist and so be inclined to favor interdisciplinary research. He may have demonstrated special abilities as a public speaker or in the writing of articles, or have a flair for teaching: such talents are indispensable in the training of lay leaders and in the use of the mass media. He may have already become expert as a consultant to other physicians, or have held positions as consultant to one or more community agencies. An excellent summary of some of the problems which are apt to face the psychiatrist who is directing a community program is furnished by Vaughan.[5]

Primarily, to re-emphasize what I have said above, he needs to regard himself as the servant before he can expect to become the master, the scientist before he assumes the right to speak with authority. Above all he needs that respect for others which is the first essential for a successful leader, whose job is to set standards and develop good team work.

Summary

The postwar recognition of mental illness as the most serious public health problem of our time has led to increasing concern at all levels of society and finally to direct intervention by the President of the United States. We are faced as a profession with the challenge of the times. Recognition of community mental health as a field in which psychoanalysts have already shown both imaginative planning and active leadership is proof that our specialty through its breadth of understanding of the causes of mental ill health and of the forces that assist or impede satisfactory interpersonal relations has invaluable potential for providing continuing leadership in meeting this great community challenge. While we must first make good our claim to some competence in defining the essentials of graduate education for psychoanalysts, the time has come when we should include in these considerations the place to be given in the curriculum for preparing psychoanalysts to take appropriate roles, if not continued active leadership, in the development and implementation of community programs for the prevention of mental illness and the care of the mentally disabled.

REFERENCES

1. Millet, J. A. P.: Education for community psychiatry. Presented May 12th, as part of Workshop on Cultural Attitudes in the Care of the Mentally Ill. Joint Meeting of Mexican Society of Neurology and Psychiatry, and the American Psychiatric Association, Mexico City, May 11-14, 1964.
2. Bellak, L.. ed.: Handbook of Community Psychiatry and Community Mental Health. New York and London, Grune & Stratton, 1964, p. 15.
3. Ibid. p. 85.
4. ——: p. 20.
5. ——: p. 405

THE ROLE OF PSYCHOANALYSIS IN COMMUNITY MENTAL HEALTH

By MAXWELL N. WEISMAN, M.D.

AS A SOCIAL PSYCHIATRIST rather than as a psychoanalyst, I am concerned with the development of a comprehensive community mental health program; nevertheless, as in all reciprocating systems, it is difficult to separate the two roles. While psychoanalysts were among the prime movers in community mental health developments in this country, they are themselves now being moved to accelerate their self-examination and self-change. The enormous ferment occasioned by the work of the planning groups, both on a grass roots and higher level, has already raised provocative questions which demand answers. Across the land one hears angry comments on the closed corporation nature of psychiatry in general and psychoanalysis in particular, on the apparent undertraining of people obviously available and potentially helpful and especially on the "overtraining" of some psychiatrists. By having labored so diligently to make psychiatry and psychoanalysis acceptable, the profession is creating a demand which it bears responsibility to satisfy.

I must also confess to a nagging sense of frustration when I consider the vast area covered by the term "community" and the shifting vagueness of the concept. Dr. Louisa Howe refers to the variety of definitions and looseness of application which could virtually eliminate any distinctive significance in the concept.[1] Definition may be based on a group's function as in a professional community of doctors or educators, or on its relationships as in a family, or on its organization as in a trade union. One usually speaks of a geographic community—a neighborhood, a suburb, a city, a region, a nation or even a community of nations. When one includes all humanity, I am reminded of the motto of the London Hospital "*Homo sum, humani nil a me alienum puto*"—"I am a man, nothing which concerns mankind do I consider foreign to me." While these words were originally uttered in Terence's play to justify the role of a gossip, they could, with equal justice, be proclaimed by the physician in his role as a community psychiatrist.

Dr. Howe wryly remarks "Perhaps the only useful application of the term is in describing the relatively self-sufficient life activities carried on by people in a remote village, comparatively untouched by the influence of the modern world." It is perhaps like the term "race" which is used on such varying levels—the human race, the white race, the German race, the race of doctors (See Webster's dictionary!)—as to be scientifically useful only in such single genotypic expressions as "a race of long-winged flies."

Nevertheless, we do have community psychiatrists in a burgeoning subspecialty of the profession. Heath, *v.s.*, in outlining the content of instruction in the residency and career training programs at Tulane University has limned the breadth and scope of community psychiatry and hinted at the nature of the skills involved. The huge dimension of such a program, the problems of measuring need and selecting priorities, of interrelating existing resources and facilities, of planning new ones, of financing, of securing and training personnel, of research in evaluating the effectiveness of community services—all these and more must be dealt with on some level stemming from the basic *raison d'etre*—helping those directly or indirectly in need of clinical services.

In practice the ramifications are widespread and are limited only by the imagination, initiative and resourcefulness of the trainee. He is suddenly thrust into the larger world outside the hospital walls or the confines of his own office. His general training as a physician has taught him how to work without other human resources, to rely on his own independent role. In fact, many physicians *do* work without such resources. But in community psychiatry the number and variety of individuals and organizations he may be seeing in a single day and with whom he may be closely related may be of staggering magnitude. In the Maryland training program in community psychiatry,[2] the trainee, in the mental health clinic in which he functions, may have contact with a public health officer, public health nurses, social workers, or psychologists, individually or in a group. He may consult with educators in the local schools—principals, teachers, school counselors; with legal and law enforcement personnel—judges, probation officers and police; with general practitioners or specialists in general hospitals; with social agency workers, private and public; with clergymen; with personnel from State mental hospitals; with members of the local mental health association, P.T.A., Parents Association for Retarded Children, Alcoholics Anonymous, civic organizations and other community groups. He may meet with labor leaders and sit at the same table with other behavioral scientists, with sociologists, social psychologists and anthropologists. All this in addition to interviewing patients and family members, children, adolescents and adults, individually or in group or family therapy. Community Psychiatry thus involves many people and will undoubtedly, as it develops, involve many more.

This is indeed a far cry from the traditional psychoanalytic model, from the one-to-one doctor-patient relationship. Dr. Heath appropriately asks how the psychiatrist, trained in the model of intensive psychotherapy, can apply the knowledge thus acquired to helping the larger numbers of patients or potential patients he encounters in a community setting. While he correctly states that the final answer is still to emerge, it may be interesting to explore his question further since we already have some data from the accumulated efforts of the pioneers referred to by Dr. Millet and others. Certainly this question will be a central issue if Dr. Millet's far-seeing recommendation on prerequisites for applicants and the very curriculum for psychoanalytic training institutes is adopted. A major challenge in community psychiatry today is how to relate it to the training institutions.

In a sensitive and perceptive paper[4] describing the impact of social psychiatry on a psychoanalytically oriented Department of Psychiatry, Rosenbaum and Zwerling point to the fact that the physician may be dominant only because of administrative fiat or the necessities of medical responsibility. When the emphasis is on "an exploration of social systems and culture and their impact on psychiatric phenomena, rather than as a type of psychiatric practice," as Drs. Redlich and Pepper[5] have succintly defined social psychiatry, the traditional, psychoanalytically trained psychiatrist feels insecure and perhaps even threatened by groups of people from different disciplines speaking a different professional language. It is quite conceivable that the psychoanalyst might not be the ideal captain of such a team. That he plays an indispensable role both as clinical physician and highly trained expert in the dynamics of psychopathology represents the only basic relationship to the other members of the group.

In a recent three-day conference entitled "The Worker and Mental Health" called jointly by the Maryland Department of Mental Hygiene and the State AFL-CIO, fifty busy labor leaders and workers from the shops, practicing psychoanalysts and psychoanalytically oriented psychiatrists, psychologists, sociologists, social workers, community leaders, etc. met to focus on the reasons for the underutilization of mental health services by the worker and on the possibility of expanding such services through mutual cooperation. Provocative papers were presented for discussion but the gold of the conference lay in the small workshop groups where a new language had to be hammered out and where, by the end of the third day, trade unionists and psychoanalysts were calling each other by their first names. It is significant that in the continuations committee, whose charge is to lay careful plans for joint functioning, a labor leader with a 7th grade education is beginning to play a dominant role.

In the second place, there are not only differences in emphasis on structure and function between social and individual psychiatry, but also in levels of complexity and integration. This does not state that the individual is less complex than society. As pointed out by Novikoff[6], the process of evolution of matter from the formation of the earth to the social changes in society is continuous because it is neverending and at the same time discontinuous because it passes through a series of different levels of organization—physical, chemical, biological, and sociological. The individual human may be considered as complex as society but the *order* of complexity and integration is different. What is a whole on one level is only a part on a higher level. The laws governing social phenomena cannot be fully understood without a knowledge of the laws of the lower level, including those of psychodynamics, but at the same time they cannot be predicted, a priori, *from* those of other levels. Their discovery involves methods of research uniquely appropriate to that level. While the social psychiatrist operates in a conceptual framework different from that of the individual psychiatrist, without a knowledge of the laws of human behavior derived from psychoanalytic insight, he cannot adequately describe and analyze the characteristics of the milieu in which he functions. At the same time, he will have to learn the qualitative uniqueness of the laws of the field within which he functions.

In the field of community psychiatry, we are concerned with the necessity of helping large numbers of people now, at once, without the professional manpower or, in fact, the patient population to permit of prolonged techniques of introspection. The goals in community psychiatry are thus more immediate than those in individual psychoanalysis. The analyst aims at the corrective emotional experience by modifying the original vision by the child of an experience he could not comprehend. The distortion had become incorporated into his present functioning. Community psychiatric programs have invested little time or effort in correcting this original vision, but rather are concerned with whatever current living experience can be altered.

A few examples may be illustrative. The Work—Experience Program of the metropolitan Baltimore Mental Health Association[7] provides for a kind of unsheltered workshop relationship of patients to the usual core of volunteers who work in the office to assist in the manifold tasks so active an organization must shoulder. The atmosphere is that of any business office; there are no incidents and the work gets done. The patients work as phone receptionists, bookkeepers, typists, file clerks, label pasters, news release writers, poster designers, etc. For every three patient-trainees, there is one volunteer who does the same kind of work, eats around the same lunch table and brings all the healthy, spontaneous reactions of the normal community into the office. Each trainee, however, is under the care of a psychiatrist elsewhere and is supervised by the office manager in weekly discussions on his performance, appearance and relationship with other workers.

One trainee was a long-term inpatient at a State mental hospital where the staff thought she was too ill to seek employment. In the office she proved to be an efficient and conscientious typist and stenographer and was soon discharged to work elsewhere. Another patient, a schizophrenic who saw herself as a baby and lived on baby food had been in a private mental hospital for four years. Her psychotherapist was dynamically oriented and worked with her during her stay in the program. Six weeks after she entered, she gave up baby food, her bizarre appearance changed and her fragile ego-identification was strengthened. In the work—experience program the time was spent in addressing herself not to the question of why she saw herself as a baby but to that of how she functioned each day and on her relations with other co-workers. In this program rehabilitation is accomplished, on the average, in 3.3 months after an average hospitalization of 24.1 months. Only 7.4 per cent of the group has been rehospitalized since the program began in 1960, in comparison with the Maryland and national readmission rate of between 35 and 40 per cent. As far as we know this is the first such program in this country to be carried out in a health-oriented community setting, under nonmedical supervision, using untrained community volunteers as co-workers.

Another program organized by the Maryland Department of Mental Hygiene took 90 chronically ill patients from four State mental hospitals to a summer camp for two weeks of modified standard camp activities. The patients were between 40 and 60 years of age and had been hospitalized from two to 32 years. They had been unable to function outside the hospital in spite of having made numerous attempts. It was very soon apparent that at camp many unanticipated changes were taking place. One schizophrenic hadn't spoken in five years. Her first words that first day at camp were "This is freedom." When the patients returned to the hospitals, forty-one of the ninety were able to leave within three months. All but two have now remained in their community for at least a year.

Much research remains to be done on what kind of psychotic patients may be reached by techniques entirely related to contemporary experience. We have currently been led to other efforts than the traditional psychoanalytic techniques for managing large numbers of the mentally ill. Having had such experiences we must evaluate the validity of these expedients and may conclude that such patients will respond to procedures which force them to react to group processes in contemporary reality. If the patient can reverse the pattern of failure in interpersonal relations and succeed, he may be more susceptible to an acceptance of insights into how he had operated in the past.

But there are other patients than this group alone, large as it may be, in and out of the hospitals. There is the challenge of the millions of alcoholics and the pathetically small numbers we influence or even reach in community mental health programs. The character disorders, the juvenile delinquents, the emotionally disturbed adolescents and children, the aging—these and many more swell the ranks of those needing our help. In the face of such need, when every conceivable resource must be examined and utilized, it is as ludicrous for skeptics to doubt that psychoanalysts can play a role as it is imperative for psychoanalysts to become involved in community psychiatry.

What can they contribute? Their exquisitely deep and sensitive knowledge of human dynamics, their very real awareness of intrapsychic forces and interpersonal relations and their many other special gifts are indispensable on *all* levels—as planners, researchers, consultants, teachers, therapists. The developments of psychoanalytic ego-psychology, when applied to the community, may transform the content of education and training. If the one-to-one patient-doctor relationship is found inappropriate as the sole technique in community programs, surely a wider application can be found for the theory and content of psychoanalysis, as well as for its use as a research tool. Psychoanalysts should, in addition, be best equipped to engage in preventive programs since they know so well what goes into people's lives which may doom them for life. Their influence in legislative pressures and social action for preventive programs may thereby be immeasurable.

In conclusion, it is worth referring to the prophetic words Freud read before the Fifth International Psychoanalytic Congress in 1918. They have often been quoted in other connections but they bear repeating as a coda to this volume:

> "I will cast a glance at a situation which belongs to the future—one that will seem fantastic to many of you, but which I think, nevertheless, deserves that we should be prepared for it in our minds. You know that our therapeutic activities are not very far-reaching. There are only a handful of us, and even by working very hard each one can devote himself in a year to only a small number of patients. Compared to the vast amount of neurotic misery which there is in this world, and perhaps need not be, the quantity we can do away with is almost negligible. Besides this, the necessities of our existence limit our work to the well-to-do classes. . . . At present we can do nothing for the wider social strata, who suffer extremely seriously from neurosis.
>
> Now let us assume that by some kind of organization we succeeded in increasing our numbers to an extent sufficient for treating a considerable mass of the population. On the other hand, it is possible to foresee that at some time or other the conscience of society will wake and remind it that the poor man should have just as much right to assistance for his mind as he now has to the life-saving help offered by surgery; and that the neuroses threaten public health no less than tuberculosis, and can be left as little as the latter to the impotent care of individual members of the community. When this happens, institutions or out-patient clinics will be started, to which analytically-trained physicians will be appointed. . . . It may be a long time before the State comes to see these duties as urgent. . . . Some time or other, however, it must come to this.
>
> We shall then be faced by the task of adapting our technique to the new conditions. . . . It is very probable, too, that the large-scale application of our therapy will compel us to alloy the pure gold of analysis freely with the copper of direct suggestion; and hypnotic influence, too, might find a place in it again, as it has in the treatment of war neuroses. But whatever form this psychotherapy for the people may take, whatever the elements out of which it is compounded, its most effective and most important ingredients will assuredly remain those borrowed from strict and untendentious psychoanalysis."[8]

The time has arrived to heed this clarion call.

REFERENCES

1. Howe, L. P.: The concept of the community: Some implications for the development of community psychiatry. *In* Handbook of Community Psychiatry and Community Mental Health. L. Bellak, ed. New York and London, Grune & Stratton, 1964.
2. Kern, Jr., H. M., Spiro, H. R., and Kolmer, M. D.: An Appraisal of the Maryland Training Program on Community Psychiatry, delivered at the American Public Health Association meeting, N.Y.C. October, 1964.
3. Lief, H. I.: What your Doctor Probably Doesn't Know about Sex. Harpers Magazine 229:92-96, 1964.
4. Rosenbaum, M., and Zwerling, I.: Impact of Social Psychiatry, Arch. Gen. Psychiat. 11:31-39, 1964.
5. Redlich, F. C., and Pepper, M. P.: Social Psychiatry. Am. J. Psychiat. 116:611-16, 1960.
6. Novikoff, A. B.: The concept of integrative levels and biology. Science 101:209-215, 1945.
7. Michelson, E.: Volunteer work for mental patients. J. Vocational Rehabilitation 30:14-16, 1964.
8. Freud, S.: Lines of Advance in Psychoanalytic Therapy. Standard edition of the Complete Psychological Works. London, Hogarth Press, vol. 17, 1955, pp. 166-168.

NAME INDEX

(Principal Reference in Italics)

SUBJECT INDEX

(Principal References in *Italics*)

Thanks are due to Mary Farkas, M.D., for aid in the preparation of this index.